Advances in Spatial Science

Editorial Board
David F. Batten
Manfred M. Fischer
Geoffrey J.D. Hewings
Peter Nijkamp
Folke Snickars (Coordinating Editor)

Springer
Berlin
Heidelberg
New York
Barcelona
Budapest
Hong Kong
London
Milan
Paris
Santa Clara
Singapore
Tokyo

Titles in the Series

Cristoforo S. Bertuglia, Manfred M. Fischer
and *Giorgio Preto* (Eds.)
Technological Change, Economic Development and Space
XVI, 354 pages. 1995

Harry Coccossis, Peter Nijkamp (Eds.)
Overcoming Isolation
Information and Transportation
Networks in Development Strategies for Peripheral Areas
VIII, 272 pages. 1995

Luc Anselin, Raymond J.G.M. Florax (Eds.)
New Directions in Spatial Econometrics
XIX, 420 pages. 1995

Heikki Eskelinen
Folke Snickars (Eds.)

Competitive European Peripheries

With 22 Figures

Springer

Dr. Heikki Eskelinen
University of Joensuu
Karelian Institute
P.O. Box 111
FIN-80101 Joensuu, Finland

Professor Dr. Folke Snickars
Royal Institute of Technology
Department of Infrastructure and Planning
S-10044 Stockholm, Sweden

Die Deutsche Bibliothek - CIP-Einheitsaufnahme

Competitive European peripheries / Heikki Eskelinen ; Folke
Snickars (ed.). - Berlin ; Heidelberg, New Yoek ; Barcelona ;
Budapest ; Hong Kong ; London ; Milan ; Paris ; Tokyo :
Springer, 1995
  (Advances in spatial science)
  ISBN 3-540-60211-9
NE: Eskelinen, Heikki [Hrsg.]

ISBN 3-540-60211-9 Springer-Verlag Berlin Heidelberg New York

This work is subject to copyright. All rights are reserved, whether the whole or part of the material is concerned, specifically the rights of translation, reprinting, reuse of illustrations, recitation, broadcasting, reproduction on microfilm or in any other way, and storage in data banks. Duplication of this publication or parts thereof is permitted only under the provisions of the German Copyright Law of September 9, 1965, in its current version, and permiss7ion for use must always be obtained from Springer-Verlag. Violations are liable for prosecution under the German Copyright Law.

© Springer-Verlag Berlin · Heidelberg 1995
Printed in Germany

The use of general descriptive names, registered names, trademarks, etc. in this publication does not imply, even in the absence of a specific statement, that such names are exempt from the relevant protective laws and regulations and therefore free for general use.

SPIN: 10510413   42/2202 - 5 4 3 2 1 0 - Printed on acid-free paper

Preface

Europe's space is in a flux. Earlier cores and peripheries in Europe are experiencing a profound transformation. The driving forces include, amongst others, Western European economic and political integration, and Eastern European transition. We are also witnessing fundamental technological and organisational restructuring of industrial systems. Information technology and telecommunications are rapidly altering the requisites for comparative advantage. Peripherality is being determined more by access to networks than by geographical location. Economies of scale can be attained in distributed networks of production with good access to markets as well as in large agglomerations. Clearly, these changes also call for new perspectives in regional analysis.

This book derives its impetus from an Advanced Summer institute in Regional Science which was arranged in Joensuu, Finland, in 1993 under the auspices of the European Regional Science Association. Some of the papers, which were discussed at the institute, were thoroughly revised for the present purpose. In addition, chapters on specific topics were specially written for the volume. In most contributions, the focus is on the Nordic countries and their internal peripheries. They form a particularly interesting case in assessing prospects for the multi-faceted centre-periphery confrontation in Europe.

The message of this book is that geographical peripherality is not necessarily a fatal syndrome, implying dependency, backwardness and isolation. Peripheral regions can, to an increasing degree, provide viable platforms for economic interaction and political cooperation. They are striving to become competitive in the new political and economic order, and they are far from being outstripped in this race. This is especially true for the Nordic countries, and for their internal peripheries. An important political challenge is the constructive reorganisation and revival of the welfare state. Also in this regard, some of the light might come from the North.

The ideas of the book have also received support from the contributors in the editorial process. Contributors have been co-operative in shaping a book with a sharp focus and a succinct message.

The final product of the effort has emerged as a result of complex interaction through different communication channels. The Karelian Institute at the University of Joensuu has provided strong support for the editorial work as has the Department of Infrastructure and Planning, at the Kungl Tekniska Högskolan, Stockholm. Professor Anton Kreukels, Department of Applied Geography and Planning, University of Utrecht is thanked for supportive action in the final phase of the editorial process. Our special thanks are due to Lea Kervinen, who edited the camera ready copy of the manuscript, and Nancy Fournier, who checked the English language.

Joensuu and Utrecht, March 1995

Heikki Eskelinen　　　　　　　　　　　　　　　　Folke Snickars

Contents

1	**Competitive European Peripheries? An Introduction** Heikki Eskelinen and Folke Snickars	1

Part I EUROPEAN SCENE   15

2	**The European Network Economy: Opportunities and Impediments** Peter Nijkamp	17
3	**Regional Development, Federalism and Interregional Co-operation** Riccardo Cappellin	41
4	**Remaking Scale: Competition and Cooperation in Prenational and Postnational Europe** Neil Smith	59
5	**Cross-border Co-operation and European Regional Policy** Anne van der Veen and Dirk-Jan Boot	75

Part II NORTHERN LIGHTS   95

6	**Europe of Regions – A Nordic View** Perttu Vartiainen and Merja Kokkonen	97
7	**Peripherality and European Integration: the Challenge Facing the Nordic Countries** Sven Illeris	115
8	**Regional Development in the Nordic Periphery** Jan Mønnesland	131
9	**The Fall and Revival of the Swedish Welfare Model: Spatial Implications** Stephen F. Fournier and Lars Olof Persson	151

10	Are Leaping Frogs Freezing? Rural Peripheries in Competition Jukka Oksa	183
11	Internationalization from the European Fringe: the Experience of SMEs Heikki Eskelinen and Leif Lindmark	205
12	The Social Construction of Peripherality: the Case of Finland and the Finnish-Russian Border Area Anssi Paasi	235
	Author Index	259
	Subject Index	263
	Contributors	269

CHAPTER 1

Competitive European Peripheries?
An Introduction

Heikki Eskelinen
University of Joensuu

Folke Snickars
Kungl Tekniska Högskolan, Stockholm

1.1 Competitive European Peripheries

Peripherality generally connotes distance, difference and dependency. A typical periphery is geographically remote, economically lagging, dependent upon external political and industrial decision-making, and culturally obsolete. In this vein, the paired antinomy "centre versus periphery" is often used to loosely characterise asymmetrical relationships and the disparities of regional systems.

Yet the centre versus periphery paradigm is not an integral part of a classificatory model used for the analyses of regional systems. It is rather, as Anssi Paasi emphasises in this volume, a contextual category, which is loaded with economic, political and cultural meanings. The centre of the world for some is the remotest periphery for others. An analogous situation may be the experience of being in the eye of a hurricane. There is no specific feeling of centrality. Rather, the turbulence is likely to arise when you wish to get away from it.

Interdependencies among regional systems and relevant actors develop over lengthy periods of time. A longitudinal approach to the study of such interactions can enable a conjoint analysis of centres and peripheries. Numerous investigations conducted through various academic disciplines have concluded that relations between the dimensions of centrality and peripherality are complex. A country or a region which is peripheral in one field can be central in another. A region which was once a periphery is not necessarily a periphery forever. At one time, Kiel was a periphery of the European Community. With the inclusion of Denmark in the European scene, the peripheral role was assumed by Aalborg. Most recently, as Sweden and Finland chose to join the EU, the periphery shifts to the outlying areas of Kiruna and Rovaniemi.

Tromsø, because of Norway's alternative decision, is not included as a peripheral region.

The title of this volume refers to peripherality as a heuristic metaphor. It has been chosen to demonstrate a challenge, which will be taken on by those regions which have been connoted as peripheral in the nation-state period in Europe, be they part of nations in continental Europe or regions in the geographical outskirts of the continent. A northern perspective is apparent in many of the contributions. The research findings enable the authors to evaluate the industrial and regional policies of the Nordic countries. A major distinction is made between small and large regional policy, that is, public policies which fall under the regional policy heading in government budgets and all public policies which have an impact on regional development.

The formation of the European Union as a political construct should at least imply a removal of the peripherality stamp on border regions between nations. In fact, the hierarchical decision-making structures in the nation-state have created gaps between regions at the borders of nations which are atypical. Borders have naturally and historically been located where the action is most intensive.

It seems obvious that one of the competitive elements of Europe in the global economy is the diversity of regional interests expressed in political, cultural or economical terms. Removing national borders offers a metaphor for lifting some political issues to the level at which public policy will have a bite in the first place. This will not lead to a unionisation in the narrow meaning of the European Union concept. Europe's historical asset is its multicultural nature. In this vibrant cultural constellation of regional systems borders will continue to be important, although the limits between nation-states will not be as easily discerned.

The individual chapters of this volume focus on the forms of peripherality and on the processes which condition centre and periphery structures. The Nordic countries and their internal peripheries are the main points of reference. These will provide a laboratory for the analysis of how the ongoing profound economical and political transformation of Europe may be reflected in the spatial division of labour and in the settlement patterns.

1.2 Centre and Periphery Paradigms Revisited

The paradigm of centre and periphery has always been imprecise with regard to spatial representation. This is disturbing since the notion makes reference to a spatial configuration concerning a geographical central spatial unit and a hinterland which is dependent on that centre. The disturbing element arises basically from the fact that continuous space is not distinguished from discrete space. The centre or the core is a node and the periphery is a territorially defined region. It is obvious that in some sense a node is always a centre because it is spatially dense, containing a concentration of economic and social activity. The notion of economies of scale is congruent with this view. On the other hand, the periphery is spatially sparse, containing small nuclei of economic and social activity. The notion of natural resources in agriculture and forestry is closely connected to this view.

In the centre and periphery literature there is seldom a convincing treatment of transportation systems and their supporting physical and organisational networks. The central node in the theory is apparently so dense that transportation is not a problem. The periphery is so sparsely settled that current theory offers little to clarify how it is to be served by transportation facilities.

An argument of this volume is that the centre and periphery discussion should be conducted in a network context. This network may most easily be conceived in spatial and physical terms, although this is by no means necessary (see e.g. Karlqvist et al. 1990). In a network of nodes and links which exhibit different configurations, and layers, addressing the question of centres and peripheries is still warranted. The answer is not so obvious. Where is the centre of a well-connected network?

The networks depicted in Figure 1.1 are given as illustrations of the complexity of the centre-periphery issue in a network context. There is no centre at all in an extended street network, as shown in the first part of the figure. The same holds for a fully connected network. A difference in the number of potential connections will arise from the delineation of the network via an external border. The border, rather than the network structure tends to induce differences in centrality, as is depicted in the middle node in the fully connected network given.

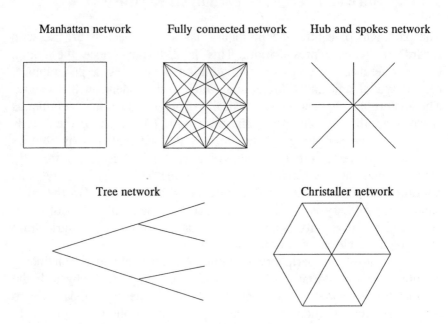

Figure 1.1. Excerpts from a sample of structured networks.

The hub-and-spokes network corresponds to the network topology of the centre and periphery idea. Centrality is clear since connections among the nodes at the end of the spokes are more effective via the hub. Modern firms with extensive reliance upon the mobility of factors which serve markets for mobility are often the hub-and-spokes type. The tree-structured network represents two different models, the hierarchical and the recursive structures. Decisions which are developing over time will logically be hierarchical since time is irreversible. Hierarchical networks are typical of public organisations and are deeply connected with the nation-state.

Finally, the hexagonal pattern of central places is of course a network. In this network a hierarchical organisation arises from the superposition of different functions to be performed in the nodes. There is no physical transportation network in the theory of central places. Transport opportunity is ubiquitous. In reality, the superposition of functions in the nodes has a correspondence in a multilayered network in which transportation, and telecommunication, can proceed simultaneously in several partly linked networks.

The opening of national borders implies that networks which were formerly contained within one nation may extend across borders. The development of this process will depend crucially on the internal organisation within a country, especially in the public sector. Regions in different nations of Europe are, to varying degrees, autonomous with respect to their national governments. Autonomous regions will have more options to create new linkages when the national hierarchies are allowed to interact directly with one another. Firms and households have been more active and independent than public organisations in this process of joining networks across national borders. Regions and nations with a tradition of trade and cultural exchange have been more instrumental than regions and nations with large internal markets for commodities and services, be they commercial, business, or cultural.

1.3 Disparities and Concentration

Obviously, the disciplinary context is of major importance in conceptualising the subordinate position of peripheries and in determining which aspects of peripherality require closer scrutiny. For instance, it is only natural that the concepts of horizontal and vertical peripherality are distinguished in policy analyses. The former contrasts one region and its population with another region. In this fashion, it avoids the multilayered nature of opportunity and dependency networks. The notion of vertical peripherality opens a different perspective to power and dependency. This concept refers to the situation of networking groups of actors both within a centre and periphery and across several centres and peripheries (see also Rokkan and Urwin 1983).

Centre-periphery patterns of the global economy have been investigated by the structuralistic school of development studies since the 1950s. Yet this approach has not gained much ground in mainstream regional analysis. In fact, the centre-periphery antinomy is usually employed without any explicit linkage to a specific paradigm in the analysis of regional disparities and locational patterns. It simply refers to the spatial concentration of activities and related disparities, leaving patterns of power and dependency aside. At the same time, it is noteworthy that the structuralist school has avoided penetrating the regional level in their analyses. The structuralist analyses of development may use global regions as their object of inquiry but less often regions of a nation in the sense of the horizontal political science perspective.

From a narrow and static perspective, peripherality boils down to the problem of accessibility. Distances give rise to transport needs which implies real costs to be borne by somebody. This disadvantage is increased further by the fact that scale economies in production and in the use of infrastructure cannot always be developed and utilised to the same extent in peripheries as in centres of where there are large numbers of potentially mobile people and numerous socio-economic activities.

Centre and periphery patterns in the location of activities derive, to a large extent, from the intrinsic characteristics of economic processes (see e.g. Krugman 1991) The spatial concentration of demand and the network nature of transport infrastructure jointly tend to sustain growth processes which have for some reason been initiated in a certain region. The resulting spatial polarisation is based essentially on increasing returns, and the process of making use of these increasing returns represents cumulative causation in the classical sense of Hirschman and Myrdal. The interdependency of demand and location will constrain regions with rigid positions in the international and interregional division of labour.

This setting of specialised regions, which dates back to the early phases of industrialisation, is clearly visible in western Europe. The centre lies around the London-Milan axis. Most geographical peripheries are also economic peripheries. Krugman (1991) speculates about the consequences of European integration through a comparison with the US experience. His conclusion is that the economic forces of production and location have led to a considerably more specialised location pattern in the US. In a future-oriented European perspective, this would mean that the London-Milan axis would internally be exhibiting clear patterns of economic specialisation. The patterns of specialisation which existed before at a lower spatial level will be repeating themselves at a broader spatial level, spurred by enhancements in transport opportunities from investments in new infrastructures.

Although the patterns of the spatial division of labour and the concentration of economic activities are relatively stable, they are not completely resistant to the effects of political upheavals, technological breakthroughs, or new infrastructures. A look to history will confirm that changes in public policy can radically transform the role of a country or region.

Not surprisingly, the possibility of dramatic changes in the prevailing centre-periphery structures has aroused a great deal of attention in Europe in recent years. Rapid technological advances, the deepening integration in western Europe and the partial opening up of eastern Europe have potentially major implications for spacio-economic dynamics (see e.g. Masser et al. 1992). The first part of this volume investigates these driving forces and ongoing changes. Peter Nijkamp and Riccardo Cappellin paint broad pictures of European transformation now and in the future. The claim for an infrastructure policy which extends over the whole of Europe is a central message of the contribution by Peter Nijkamp. He provides check-lists of problems to reflect on, and suggests approaches for tackling these problems. His chapter represents an effort to provide new analytical categories for the thinking which must precede policy-formation in the future European arena, by suggesting typologies of network-related phenomena.

In Chapter 3, Riccardo Cappellin outlines and evaluates the Europe of Regions view with a focus on institutions and regional policies. He asserts that policies based on the endogenous development or network paradigms precondition new institutional solutions to the relations between nations and sub-national jurisdictions. Interregional co-operation is an essential element of an emerging model of regional policies. It opens new prospects for peripheries which have been characterised by a dependency syndrome.

In the next chapter, Neil Smith challenges the vision that there is a harmonious utopian Europe in the making, by means of all-embracing networks and co-operation. His key concept is geographical scale, defined as the spatial resolution of contradictory social forces, in particular the resolution of opposing forces of competition and co-operation. In his contribution, Smith emphasises the conflicting elements of the ongoing transformation in Europe, including postnational processes in the west and prenational in the east. These forces also precondition the formation of potential coalitions which struggle for power in the decisions concerning the agenda of integration, and the prospects for peripheries.

In chapter 5, Anne van der Veen and Dirk-Jan Boot introduce the issue of border regions which have been typical peripheries during the period of European nation-states. They analyse the institutional systems of the countries sharing the borders, and the division of labour between regions, nations and European agencies in the border context. Their view is that the border regions will be able to create a favourable position for themselves if they are able to create a common institutional framework and use decision-systems in a flexible way. Obviously, these tasks are especially demanding on the external borders of the EU which provide a real testing ground for the ongoing integration process.

1.4   Competitive Peripheries - Northern Style

The second part of the volume focuses on recent Nordic research which bears upon the question of competitive environments and decision structures in the north of Europe. The Nordic countries, Denmark, Finland, Iceland, Norway and Sweden, provide an especially interesting case for the analysis of the complex challenges posed by the current economic, social, political, and spatial, restructuring in Europe. Sweden and Finland have recently taken historical steps to alter their political relationships with western Europe by joining the European Union. Norway has decided to stay outside of the union while Denmark has been a member for many years already. Iceland still has an independent role in the European arena. The Danish membership of the European Community has meant that there has been a Nordic division in the political relationships with other countries in western Europe for a long time. The division has extended to security policy with the countries pursuing different defence strategies. Yet the Nordic countries have very strong economic and cultural connections with one another.

As elsewhere in the western world, industrialisation has had a critical role in the formation of the prevailing spatial division of labour in the Nordic countries. Their original base of specialisation in industrial production and foreign trade was laid on natural resources; the forest resources (Finland and Sweden), hydroelectric power and ores (Norway and Sweden), agricultural land (Denmark) and fish (Iceland and Norway). The interdependencies between natural resource base, industrial structure and settlement pattern are still clearly discernible in these countries.

The export industries, which initiated the industrialisation process, were originally located where raw material, energy supplies and transport routes were plentiful, primarily in coastal areas and along inland waterways. Import activities and the bulk of footloose manufacturing production was concentrated in the central regions. The resulting geographically dispersed settlement pattern was consolidated by the construction of railway and road networks, which provided the infrastructure base for the integration of the national territories.

The distinctive economic roles of settlements and regions have obviously left their marks on the structural characteristics of the centres and peripheries. Yet there have also been powerful forces counteracting this differentiation process. Regional policy measures have been utilised to redistribute economic activities toward more remote areas, and to preserve agricultural production. The construction of the welfare state has been an even more important factor. Public services have been provided according to national norms that disregard where people live. This policy has led to improvements in the physical and social infrastructure of peripheral regions and cushioned a spatially unbalanced process of economic restructuring. On the other hand, it has led to a situation in which more peripheral regions are dependent upon jobs and income transfers by the state. Furthermore, it has led to a distributed network of public infrastructure which extends relatively uniformly to all segments of the territory.

Overall, the Nordic countries do not fit well into the stereotypic mould of peripherality. Granted, they are generally sparsely populated and geographically remote from the core regions of western Europe. Yet, according to the conventional socio-economic indicators, they have attained approximately the same level of welfare as the centre of Europe in recent decades. When they enter the European political scene, they do so with a Nordic style which has proved to be globally competitive in large segments of the industrial sector. In this sense, the recent addition of the North to the European Union has clearly strengthened the competitive edge of the European peripheries.

Explanations for the relatively successful socio-economic performance of the Nordic countries often refer to causes such as abundant natural resources, well-developed physical and social infrastructure, traditions of consensus-based decision-making and long-standing commitment to free trade. Presently, however, the strength of these factors has seemingly diminished as acute crises challenge the Nordic model of society. In the more seriously affected countries, Finland and Sweden, the symptoms include staggering economic growth and dramatically increased unemployment. These experiences have raised the question of whether the Nordic countries can successfully adapt themselves to the ongoing changes in institutions and competitive conditions. Will these peripheries maintain their current dynamics and will they continue to be regarded as positive exceptions among the fringes of western Europe?

In Chapter 6, Perttu Vartiainen and Merja Kokkonen construct a bridge between the Europe of Regions debate and the challenges faced by the Nordic countries. Illustrations of the problems and prospects facing the extended European scene are presented which support their argument that there is a tension between the Scylla of weakening national welfare systems and the Charybdis of self-reliance. Resolution can only be achieved through development strategies which utilise spatial scales both in the West and East European contexts.

A strong impetus for producing this volume is our view that the northern style of addressing issues of centre and periphery is worthy of European exposure. The Nordic countries have utilised regional development strategies that are novel in many respects. For instance, Sweden has been a widely-known point of reference in the construction of the welfare state. In this volume, authors Sven Illeris, Jan Mønnesland, Lars-Olof Persson and Stephen Fournier, and Jukka Oksa (Chapters 7–10) survey the complex constellations of central and peripheral features within the Nordic context. Their contributions present central political aspects of the welfare state in Denmark, Sweden, Finland and Norway, respectively. They also review some of the regional policies directed toward socio-economic development. Their concerns about the possible deterioration of the Nordic welfare model mirror those of policy-makers and researchers. The danger comes less from weaknesses within the system as such, than from the fundamental changes in the global economic system. These altered conditions tend to make peripherality materialise as a result of poor regional development strategies, rather than following directly from location.

It goes without saying that the issue of a feasible development strategy is far from being settled in the current Nordic debate, as it examines possible repercussions of the deepening integration. Norway, as an outlier from the EU, is an especially interesting case in this respect. The contribution by Jan Mønnesland, which has a strong flavour of the Norwegian anti-EU campaign, therefore reflects this effectively.

1.5 Structural Transformation and Institutional Transition

The modern image of the Nordic peripheries is due, in large part, to the potential of their natural resources, the egalitarian philosophy underlying the construction of the welfare state, and successful regional policies. Yet, as already suggested, the prospects for their future socio-economic development may be questioned on several grounds. In brief, they face both a traditional and an emergent peripherality problem. Furthermore, the ongoing major institutional changes in Europe are of considerable importance with regard to their economic dynamics and spatial structure.

Physical peripherality results from the lack of accessibility. Although diminishing transport costs and the development of information technologies have certainly decreased the potential isolation of remote regions, they have not abolished it completely. The accessibility problem might actually increase as some new infrastructure technologies are not economically viable in conditions of scattered demand. In addition, mega-projects, such as the trans-European networks, are confined to central regions (see Spiekermann & Wegener 1994). The traditional peripherality syndrome is also clearly relevant at a local level, where the restructuring of socio-economic processes is spatially differentiated, and conditioned by local proximity to interregional access nodes. This is analysed from a political-sociological perspective by Jukka Oksa in the present volume. He shows how strategic thinking in a new social context substantially effects the ability to link into the global economy.

The emerging structural problem for the Nordic peripheries has arisen from basic trends in international competition (see e.g. Amin & Malmberg 1993). Given that competitive advantage is increasingly achieved through man-made efforts, creativity, learning and innovation are critically important. From this dynamic perspective, it can be argued that the most important obstacle to the structural adaptation of the Nordic peripheries is not physical accessibility, but weaknesses in their operational environment (see e.g. Storper 1995).

In Chapter 11 Heikki Eskelinen and Leif Lindmark evaluate changes in strategic competitive factors and the role of local milieus in the process of internationalisation of small and medium-sized enterprises. They examine empirically whether small- and medium-sized enterprises have the resources to cope with internationalisation within different regions of the Nordic countries. The ongoing political and institutional metamorphosis in Europe tends to boost autonomous changes in competitive factors. In addition, it opens perspectives for developing new economic and spatial network structures.

The Nordic countries have historically gravitated toward western Europe (see Peschel 1993). The integration process within the European Union, in which Finland and Sweden are participating as members from the beginning of 1995, and to which Norway is linked through the agreement on the European Economic Area, gives them new opportunities to reinforce these age-old connections with the core regions of western Europe. At the same time, the more homogenous institutional setting might make the peripheries more vulnerable to competition.

In addition to strengthening their links with Western Europe, the Nordic countries are currently developing co-operation with their neighbouring regions and countries in eastern Europe. This is part of the interaction through which the sphere of influence of the former Soviet Union will be connected with the regions of western Europe. Obviously, peripheral border regions have a potentially significant role as intermediators in this process. Its advancement is in practice conditioned by political preconditions for East-West co-operation in Europe and also by the amount and nature of strategic infrastructure investments which will be made on the borders of the great European divide. The Nordic peripheries are at a crossroads in these respects as well.

Peripherality in the context of institutional transformation is analysed in this volume by Anssi Paasi (Chapter 12). He uses the border between Finland and Russia as a point of reference in his scrutiny of the changing representations of the national border in a long-term historical perspective.

1.6 Peripheries as Laboratories for Dynamic Competition

Traditionally, regions have been conceptualised as geographical places or projections of socio-economic processes. For this reason, their status has been relatively low in economic and social theories. In these theories a network paradigm has normally been at least implicitly present. This setting is rapidly changing as a result of new developments in theories of technological change and economic growth. Regions are increasingly seen as important objects of research in analyses of the endogenous competitive dynamics of contemporary socio-economic development.

The analysis of social and economic processes calls for new directions. The Nordic peripheries provide an interesting laboratory for assessing the structural adaptation and strategies regarding non-metropolitan areas at large. Basic mechanisms such as the utilisation of new material and human resources, the socio-economic restructuring, the redesign of the local and regional political system, and the potentials of new infrastructure networks may all trigger development. These are most clearly discernible in these peripheral regions. The network mesh may then be disentangled, understood, and managed.

The settlement system in peripheral and rural regions has traditionally been based on the core and hinterland model. Raw materials from rural areas have been processed in local centres and exported. This model, however, is losing its intellectual and economic base. Centres in sparsely populated peripheries would be able to act as nodes in a global economy. In this setting, peripherality is no longer linked with continuous space and centre and hinterland patterns. Contacts would reach any destination according to the needs of economic actors. In this situation, a position as a node has to be acquired or created, managed and sustained. The position is not automatically retained even if there is both a raw material supply and a demand for private and public services in the surrounding areas.

As already noted, the present structure of the northern peripheries is largely a result of the activities of the public sector. The quality of life there is the ultimate pride of the Nordic welfare system. The ongoing transformation mainly involves the forms and resources of this public intervention. The prevailing spatial structure provides development strategies with their own constraints. The expansion of urban regions by means of new transport networks is limited in geographically remote regions with low population densities. Educational opportunities cannot be evenly spread over the settlement system, at least not without access to telecommunications. This situation emphasises the significance of improving multifaceted accessibility and the necessity of innovative efforts for upgrading and developing human capital. In these respects the recipe for the people in the Nordic peripheries will not be much different from the recipe for the continental Europeans.

References

Amin, A. & Malmberg, A. (1992), 'Competing Structural and Institutional Influences on the Geography of Production in Europe', *Environment & Planning* A, 24, 401–416.

Karlqvist, A. (ed) (1990), *Nätverk. Begrepp och tillämpningar i samhällsvetenskapen*, Gidlunds, Värnamo.

Krugman, P. (1991), *Geography and Trade*, The MIT Press, Cambridge, Mass.

Masser, I., Svidén, O. & Wegener, M. (1992), *The Geography of Europe's Futures*, Belhaven Press, London and New York.

Peschel, K. (1993), *The Nordic Area in Europe's Future*, Diskussionsbeiträge aus dem Institut fur Regionalforschung der Universität Kiel Nr 26/1993.

Rokkan, S. & Urwin, D.W. (1983), *The Politics of Territorial Identity*, Sage. Beverly Hills.

Spiekermann, K., and Wegener, M. (1994), 'The shrinking continent: new time-space maps of Europe', *Environment and Planning B: Planning and Design*, 21, 653-673.

Storper, M. (1995), 'The resurgence of regional economies: the region as a nexus of untraded interdependencies', *European Urban and Regional Studies*, 2 (3), (forthcoming).

Part I
EUROPEAN SCENE

CHAPTER 2

The European Network Economy: Opportunities and Impediments

Peter Nijkamp
Free University, Amsterdam

2.1 An Outsider's View of Europe: Preface

Economic integration, political diversity and socio-cultural identity are the current confusing features of Europe. Uniformity and heterogeneity seem to run parallel in Europe's race towards a new profile in the international arena.

In order to offer some reflections on the future of Europe after its integration and on the emerging European network economy, it may be interesting to start with a few citations from four well-known American economists who participated in a panel discussion at the 67th Annual Conference of the Western Economic Association International (July 10, 1992, San Francisco). They were published as a special contribution in *Contemporary Policy Issues* (11(1993): 2, 1-22).

> I think what we really are interested in here today is that the world in some ways is emerging and that its direction is uncertain in terms of international political structure. One potential direction is a movement away from the uniform community of nations – with at least legally and juridically the same rank – toward bigger entities, of which the European Community is the most conspicuous. Of course, we have moves in the opposite direction, such as in the case of the Soviet Union. But the circumstances there are somewhat special, so the general direction is not at all mixed.
>
> (K.J. Arrow, 2)

The result of these transcendent influences – transcendent in the sense of transcending the nation-state – is to move toward globalization of business, professional, and social groups. As a general tendency, I think multinational regionalization – be it EC or NAFTA or OPEC – is a politically managed, if not manipulated, interim stage in the context of a sort of general move toward transcendence of the nation-state. In sum, there are strong technological forces tending to transcend the nation-state.

(C. Wolf Jr., 7)

After some 40 years of economic integration, particularly in Europe, this process may change as a result of changes in the world situation. Many have come to regard economic integration as inevitable, as continuing dynamic process. But in recent years, a new dynamic process has emerged, that of disintegration. Thus, competing dynamic processes whose resolution could lead to a shift in the prevailing regime have emerged. The emergence of these new dynamic processes stems from the structural break that occurred in the international system over the period since 1989, involving the revolutions in central and eastern Europe, the end of the Cold War and of the Warsaw Pact, the unification of Germany, and the dissolution of the Soviet Union, all of which have had profound effects worldwide.

(M.D. Intrilligator, 8)

If we look at economic integration, I suppose nobody in this room opposes free trade. The problem with economic integration if it goes beyond free trade is that it develops into what we have in Europe. Regarding Schumann, by the way, I think that the United States' pressure came earlier. I think the United States wanted Europe united almost simply because we were united. The American statesmen could not imagine why Europe was not one country. So, I think we got in earlier. But basically, the Schumann plan was a cartel for coal, steel, and iron. Brussels has been busily engaged in organizing cartels ever since.

Needless to say, the agricultural program is the worst, but they all are very bad. What we would like is free trade without economic integration beyond that level, except for a few harmless things.

(G. Tullock, 15)

The above views point – despite their diversity – towards various common European developments as observed by non-Europeans. The most intriguing ones are:

- integration benefits are only to be expected in case of free trade among regions and nation-states; this may be at odds with structural protectionism (including structural subsidies to privileged sectors);
- the era of the nation-state will likely go through a dramatic change in the near future, as it will increasingly be at odds with European thinking on the one hand and the drive towards regional autonomy on the other hand;
- economic connections between groups of actors will be based less upon traditional intra-nation linkages, and much more upon trans-border network configurations driven by economic forces in which regions play a dominant role;
- the concept of 'fortress Europe' as a strong economic and political power block is far from reality in light of the internal fights among nation-states to acquire a maximum share of the European 'pie'.

Nevertheless, it has to be recognised that the recent history of Europe – despite criticisms on the Maastricht Treaty – mirrors an unprecedented dynamics in which not only the EC12 countries, but also the EFTA countries and increasingly the East-European countries are involved (see Nijkamp 1993b). Europe is evolving into a network society with a myriad of nodal centres and regions linked by infrastructure connections of different quality (Hall 1993). The links to the Nordic countries, the East-European countries and the Mediterranean countries leave much to be desired, as in many cases we do not only observe missing links but also missing networks (see Nijkamp et al. 1994). It has also become clear that a network society generates a window of opportunities for new operators who are able to benefit from a multimodal infrastructure which emphasised complementary and competing networks. This means that the integration benefits of a new network economy are not only shaped by infrastructure policies of public decision-makers but also by creative decisions of network operators who are able to combine the strong and weak points of the emerging European network society.

2.2 The Emerging European Network Society

As mentioned in the preceding section, Europe is gradually but steadily moving towards a *network society*, characterised by economic integration, political co-ordination, regional autonomy and mobility of people. Networks connect people and places and are able to generate socio-economic added value through synergy and interaction. Such networks may be physical, non-material, organisational or club-oriented in nature, exhibiting a wide spectrum of multi-layer configurations.

Networks have mainly evolved on a local and national scale with standards varying according to local policies and requirements. Recently, lack of compatibility and capacity of systems have become key constraints to future development. This issue is even a problem where there is no physical infrastructure (e.g., airlines). Control systems are needed to maximise capacity (e.g., air traffic control, rail signalling and road traffic control). Even the infrastructure itself requires heavy new investment, as large parts of the road and rail systems were constructed 50 and 100 years ago. They badly need substantial upgrading or replacement. In addition, new networks are required in peripheral regions to assist development objectives. Physical barriers have to be overcome (e.g., the Alps and the Channel), and the implementation of new infrastructure in East Europe requires large investment sums (see Figure 2.1). New technology has promoted satellite and fibre optic networks for communications. The reductions in costs of computing and networking have allowed "real time" decisions to be made. Huge data bases are available to assist in the decisions of many businesses. The move to the post-industrial society has revolutionised the ways in which existing networks are used and created opportunities for new forms of communications through city networking, data exchange and research networking (Knowles 1993).

The main feature of networks is *actor dependency* through physical and non-physical interaction (Kamann and Nijkamp 1991). Naturally, networks need an intelligent technological architecture, but their potential is largely determined by clever human decisions (see Capello 1993), so that social sciences have a clear role in network analysis. *Social science issues* in a network society concern in particular the following items:
- genesis and design of networks in the economy, in space and in social organisations and communities;
- control and decision-making mechanisms in a democratic network society;

- critical success factors for well-functioning networks in relation to economic, financial, environmental and organisational impediments;
- social, economic and technological niche-formation in networks, through which interest groups can build up a competitive advantage;
- barriers in open networks including their externalities (for instance, safety considerations in congested road networks).

Figure 2.1. Potential and strategic corridors in East and Central Europe
Source: EC (1992, 60)

It is thus noteworthy that attempts towards improving the physical network infrastructure may not be sufficient to overcome the more divisive non-physical barriers between countries and regions such as language or cultural barriers based on tradition or historical heritage. In particular, the unification of Germany and the recent opening of the borders to eastern Europe have demonstrated that bridging these non-physical gaps can take much longer than the re-integration of transport and communication networks, even though this alone may require decades. Thus there is apparently tension between potentiality and bottlenecks in the European restructuring process.

The emerging European network is not uniform, even nor equally accessible, but is characterised by a *dialectic between integration and disintegration*, which might erode the integration benefits. Examples are:
- political conditions, notably economic openness vs. closure caused by civil wars;
- socio-psychological motives, in particular a sense of locality vs. European citizenship;
- institutional considerations, characterised by the decline of the nation-state vs. the strive for regional independence;
- industrial interests, asking simultaneously for free competition vs. European protectionism.

These observations call for due attention to be given to network behaviour. As a research theme, networks are not unfamiliar territory. There exist approaches in mathematical topology, electrical engineering, hydrology, transport and information science and operations research which treat networks as directed graphs with fixed capacity carrying flows of different speed, intensity and direction. However, these network concepts are too limited to capture the complexity of interaction between network flows and the behaviour of the human actors operating and communicating over them. Only recently have there been attempts in sociology and political science to address these richer but much less tangible issues. The merging of the above two directions of thought (which may be provisionally termed 'hard' network theories and 'soft' network theories) seems to be a timely research agenda of extreme relevance to decision-making in Europe.

Emerging research questions on European networks are inter alia:
- do networks act as platforms creating conditions for more stability in a socio-economic and political sense?
- will the concept of 'fortress Europe' be reinforced and revitalised through a network configuration?

- are the social external costs of a mobile European economy compatible with the external social benefits gained?
- do sudden political transformations, such as the German re-unification, lead to excessive costs which may erode public support for the new European spirit?

The European network society is expected to exhibit unprecedented degrees of *spatial mobility* (see Nijkamp 1993a). This holds for both residential moves, industrial relocation and international migration. Areas which need clarification are inter alia:
- explanatory backgrounds of a mobile society, at both a micro and meso/macro level;
- implications of a 'declining European space' and of the loss of the 'home of man', as reflected in the rapid rise of regional autonomy ideas;
- multidisciplinary analysis and prediction of international migration in the European space, not only in relation to guest workers but also in relation to political refugees (Europe as an immigration continent);
- relationships between a mobile network society and national to local environmental externalities, especially from the viewpoint of global environmental change;
- the role of new technologies as telematics for the behaviour of people in a dynamic European network space and the design of new maps of Europe that might emerge (see e.g., Masser et al. 1992).

It suffices to say that the construction of a network society does not materialise automatically, but requires dedicated efforts from both public and private sectors. Substantial capital investment is required to construct high quality networks and difficult decisions must be made for the European dimension to be considered as important as the national concerns. Traditionally, transport infrastructure investment has been carried out by national governments in the public sector. It is only in the communications sector that the possibility of private capital input has been explored. New European agencies (e.g., EBRD and EIB) have been set up to adjudicate on new investments, and possibilities are also being explored of joint venture projects between the private and the public sectors. In the operations of transport and communications markets, many European countries have had different traditions, some based on strong central intervention and others allowing much greater market freedom. Under these different political regimes, networks evolve in different ways. For example, with respect to bus and air transport in a deregulated market the structure moves from a comprehensive network of services with many links to one based on a hub and spoke configuration with longer distances to be travelled, but with more frequent services. This structure results in significant savings to the operator and a less accessible market.

In the context of regulatory policy on networks the role of governments is of utmost importance. Most decisions regarding European networks are taken by national governments through well-established procedures. As transnational European networks evolve, many decisions will have to be taken by international agencies. This requires the establishment of new institutional, organisational and legal frameworks. The roles of the different political, legal, financial and planning agencies must be determined together with an understanding of the process of decisionmaking. The implications of decisions taken at one level in the process must be accommodated at other levels of integration if equity and efficiency are to be maintained. The current debate on subsidiarity within the EU and the appropriate form of Environmental Impact Assessment in transport investment decisions is a good example of the problems raised. In addition to the EU political dimension, there are important issues of harmonisation and standardisation in networks, access to information, the organisational culture of networks as well as institutional and organisational barriers linked to networks.

2.3 The Future Challenges of European Transport

Networks generate synergy through (physical and non-physical) spatial interaction. Transportation fulfils a key role in modern societies, not only for road users, but also for many other actors: public authorities, network operators, industry and society at large. In the same vein, transport is assuming a central role in the new European field. The context and nature of European trade and transport is entering a new era. As mentioned previously, Europe now offers a scene of dramatic change: integration of the EU market, disintegration of various nation states, and more openness between all countries and regions in Europe.

From a global perspective, traditional patterns of competition – within national borders – are increasingly being replaced by vigorous competition on a multi-national and even world-wide scale. "Intra-country" competition is being replaced by "inter-trade-block" competition as traditional boundaries disappear. This is already taking place in Europe and will take place in other parts of the world as well. Countries within such trade-blocks are becoming part of an open economic network. To maximise the competitiveness of such a network, and thereby maximise its socio-economic potential and performance, attention must be paid to the quality of its transport infrastructure. Transport has become an important component of modern production processes, among other things because of intensified division of tasks between firms (in different countries) and the logistic integration of business processes. At the same time, large metropolitan areas appear to become poles of competition in an international context, so that the quality of a metropolitan network will also play a pivotal role.

As a result of globalization and the rapid rise in international interaction and communication, transportation in Europe (both passengers and freight) has grown enormously, especially in recent years. As the supply of infrastructure has followed this trend only in part, existing infrastructure bottlenecks have become accentuated. This is a very serious problem, since economic development and infrastructure development have always been strongly interlinked, as is shown by hundreds of years of European history. The full benefits of the foreseen Internal European Market will only be reaped in case of effective (physical and non-physical) infrastructure adjustments in Europe. What is needed in this context, is European – and *not* national – thinking and action in infrastructure policy, based on knowledge of past successes and failures in infrastructure planning and of the future needs of the economy, the people living in Europe and their (endangered natural) environment. Not only in the field of passenger transport, but notably in the field of freight transport, networks in Europe are not performing at a competitive scale. This applies for all six basic networks: rail, combined transport, road, inland waterways, airports and seaports.

Furthermore, the structure of production, distribution and transport has gone through a rapid transition phase. Integrated logistics inside firms are increasingly linked to external distributional and market logistics, a tendency which leads inter alia to *logistic platforms* in an international network in order to fulfil the needs of just in time (JIT) delivery and material requirements planning (MRP). Multimodal transport will play a critical role in this new development, as is also witnessed in recent policy documents of the Commission, e.g. in the framework of the EURET programme.

The trend towards globalisation (or at least internationalisation) and the need for more competition at all levels in the new European setting have provoked a profound interest in the functioning of networks in Europe. Traditionally, the interest in networks was instigated by supply side motives, but it is increasingly recognised that the new competitive behaviour of firms in Europe requires us to focus much more directly on those actors who co-ordinate, manage and operate flows in this network. Consequently, much more attention is needed for *demand driven* activities in the transport sector.

Unfortunately, widespread interest in a *European* orientation of users and organisers of transport in cross-border networks is a very recent development, since transport policy and planning were seldom performed at this scale. National frontiers have always provided clear physical and institutional barriers between countries, although the creative behaviour of network actors has produced increased transport demand in Europe. Intra-European transport infrastructure networks have not followed this rising trend in international mobility and now exhibit bottlenecks in terms of *missing links* and *missing networks*. The emerging Internal Market among the members of the European Community has focused European politicians and industry (in a more pronounced way) on issues of socio-economic harmonisation in order to remove distortions to free competition between industries in its member states. As a result increasing consideration is now given to transportation. The Maastricht Treaty has reinforced the critical function of transportation (infrastructure) for economic cohesion in Europe. But the way towards real value added networks based on *interoperability*, *interconnectivity* and *integrated chains* is still fraught with obstacles, as it also requires a focus on competitive actors in the transport market.

Consequently, a new element to be considered in the current European transport policy scene is the changing role of actors in this field, in both the public domain (e.g., infrastructure owners or transport authorities) and the private domain (e.g., freight forwarders or logistics suppliers). A major issue is whether and how transport regulatory policy can be used to create conditions for fair competition, based on a creative division of tasks between public authorities and private actors with the goal of generating added value on using intermodal European networks.

Clearly, economic development and infrastructure development reinforce one another. Therefore, the European economy remains critically dependent upon *well functioning core networks as catalysts for future development*. Networks are a vehicle for indigenous development. Presently, however, there is a growing awareness that the current European infrastructure system is becoming outdated, without being sufficiently upgraded or replaced by modern facilities which would give the European economies a competitive edge. *Missing networks* emerge because transportation systems are developed in a segmented way, each country seeking its own solution for each transport mode without keeping an eye on the synergetic effects of a co-ordinated design and use of advanced infrastructures by various actors. Another reason for missing networks is the focus on hardware and the neglect of software and organisational aspects as well as financial and ecological implications. Cabotage, protection of national carriers, segmented European railway companies, and lack of multi-modal transport strategies are but a few indications of low performing European networks.

A European orientation towards the needs and behaviour of key actors in the integration of transport modes is necessary to cope with the current problems of missing and competing networks. It is therefore of great importance that the idea of Trans-European Networks is nowadays strongly advocated by the European Commission. It is equally important that the strategic position of public and private actors (suppliers and users) is better understood in network policy. Creative use of multi-modal networks may transform competition into complementarity and better ensure sustainable transport.

Thus, the future of a unified Europe is critically dependent upon the functioning of strategic infrastructure networks which are *interconnected* in terms of (1) *integration* between different layers of a network (e.g., co-ordination of high speed/long distance networks such as TGV or airplane and lower speed local networks such as light rail or roads), and (2) *intermodality* between different competing or complementary network modalities. In this respect also the quality of *nodal centres* (terminals, stations, urban centres) plays an important role, as well as the magnitudes of different types of transport (or carriers) in Europe.

The notion of *interoperability* of networks, as advocated in the Maastricht Treaty, generates an additional series of important issues which deserve thorough attention by policy-makers and researchers:

- the close connection between the development of transport networks and (tele)communication networks (including new logistical systems) and their potential implications for the European space (e.g., polarisation tendencies towards larger metropolitan areas);
- the emerging conflict between environmental sustainability, infrastructure expansion and competing networks (notably competing transport modes);
- the lack of standardisation of transport systems technologies in Europe, which hamper the full benefits of an interoperable European network;
- the completely different financing regimes for European transport modes, which prevent a fair competition;
- the lack of strategic insight into the linkage between European networks and global networks developed in other regions outside Europe, including the behaviour of 'network actors' who aim to fulfil the needs of a global economy.

Consequently, the policy agenda for interoperable European networks is vast (Capineri 1993) and deserves much attention in the near future, with a particular view on integrating network operators.

2.4 The Role of Network Actors

International competitiveness is a necessary condition for enhancing the level of European economic performance after the completion of the internal market. Segmented and nationalistic infrastructure policy may at best serve the short-run interests of infrastructure owners. In the long run it is a detriment to all network owners (and users) and affects Europe's economic position. Thus transportation and communication policies require a balanced implementation of actions which ensure consideration of both private and social costs, and a global orientation which exceeds country-based or segmented policy strategies. The current plans regarding the European high-speed railway system are a clear case of creative action-oriented policy analysis, even though the technology policy underlying this system mainly serves the interests of individual countries.

Networks are at the same time objects through which nations (or regions) can control part of the international (or interregional) competition. Monopolistic and oligopolistic structures in space are the result. The socio-economic benefits of co-ordination and harmonisation are often neglected in favour of emphasis on nationalistic interest. Much new research is needed on the economic importance of the existence of (deliberate and coincidental) barriers in international networks (including the missing links and missing networks phenomena).

Although in recent years the attention at both national and European levels has been increasingly focused on Trans-European Networks, it turns out that for the time being the actual interest is towards separate, i.e. single mode, transport solutions. Only recently, the awareness is growing that interconnected networks (supported inter alia by modern telecommunications and information technology) may offer a high added value. Despite its potential, *interoperability* between different modes with a view on *cohesion* of European transport systems in order to use the transport capacity as efficiently as possible appears to be very difficult to achieve in practice. Two factors of strategic importance have to be envisaged in this context: *complementarity* between different nodes in order to benefit from synergy in terms of added value networks, and *competition* between different nodes in order to operate under the most cost-efficient conditions at a European scale.

It is clear that the goal to maximise value added from the use and operation of a multimodal international network will best be reached if the impediments to free access of networks are at a minimum. Only reasons of socio-economic distributive impacts may temporarily restrict free entry, but efficiency through competition is normally best served when actors are given a free choice of different modes. This means that integration benefits will be higher since third parties are able to reap the advantages of an interconnected infrastructure network. This once more emphasises the need to look into the behaviour of key actors as a foundation for international network policy. Adequate attention is needed for the integration functions of new actors and operators in the transport market.

Transportation planning is often associated with physical movement, with infrastructure configurations and with regulations. Far less attention is paid to the way the transport market is organised, and how this organisation uses and shapes transport modalities. Especially the transaction theory of firms has shed new light on the interesting link between firm behaviour and network development (e.g., hub and spokes systems). Even though transport systems appear as fragmented networks, various operators (e.g., forwarding agencies, logistics suppliers) through multi-modal shipping, integral logistics and neo-fordist customised delivery are able to exploit transport networks for generating added value, not only in a local-regional but also in an international context. Globalisation of markets, new forms of competition, improved client orientation, integration of production and warehousing, and transport innovations are shaping new opportunities for creative actors in the transport market. These are reflected in joint ventures, 'filières', and vertical integration (see Figure 2.2).

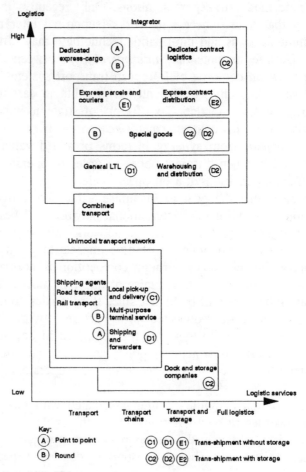

Figure 2.2. Types of logistics services providers
Source: OECD (1992, 89)

These new operators may to a large extent be considered as integrating actors in a spatial transport system which can be typified according to:
- the structure of the transport market (free competition, regulated market),
- the type of mode (road, rail, waterways, air),
- the geographical coverage (local, global),
- the quality of service (scale, scope, tariff system),
- the sophistication of transportation technology (logistic platforms, telematics, information systems),
- the structure of the network (hierarchy, hub and spokes),
- the territorial and modal policy competence on networks,
- the barriers to a full performance of networks (regulations, conflict of competence),
- the integration with telecommunication (EDI).

In the current European setting three related terms are often used: interoperability, interconnectivity and intermodality. *Interoperability* refers mainly to operational and technical uniformity which allow actors and operators to use and link various layers or components of a transport network. *Interconnectivity* is in particular concerned with horizontal co-ordination and access to networks of a different geographical coverage. Finally, *intermodality* addresses the issue of the sequential use of different transport modes in the chain of transport.

The current popularity of network concepts is undeniably connected with the declining domain of public policy: networks tend to become the channels through which competition flourishes. Both external megatrends and internal system's forces necessitate a market orientation paralleled by risk minimisation strategies. Networks seem to offer more certainty in terms of expected consequences of strategic decisions and hence may be regarded as a critical success factor in (inter)national competition.

The set of network policy actions that can be envisaged is vast and ranges from direct public supply or intervention to user charge principles or complete laissez-faire. A major challenge of network owners and operators will be to formulate strategic plans that convincingly incorporate non-zero-sum game strategies with gains for all parties involved. This may be illustrated by means of some examples.

The 'user charge' principle in transport policy has in particular become a success in those countries where suppliers and users of transport infrastructure were all enjoying benefits (e.g., suppliers by receiving more revenues from road charges, users by increasing their travel speed etc.). Likewise the question of intermodal substitution (e.g., from the car or lorry to the train) will critically depend on the willingness to implement such incentives.

A subsequent issue – and probably the most difficult one – is the design of an *assessment/evaluation methodology* for transport network policy-making. This would have to be based on *performance indicators* for both private and public actors:

- productivity gains or added value,
- network synergy based on public service delivery to private and public actors,
- competitive improvement for firms,
- spatial-territorial integration,
- technological harmonisation,
- removal of bottlenecks or spatial externalities (e.g., congestion, environmental stress, road fatalities),
- user possibilities by various specific groups (e.g., small and medium size enterprises),
- financial costs/revenues for public, (semi-) public or private bodies in charge of operating the infrastructure,
- contribution to European cohesion,
- access and benefits for less favoured regions,
- intermodal complementarity,
- degree of interoperability,
- use of telecommunication technologies (e.g., informatics, telematics).

The previous considerations on an evaluation methodology can be combined in a comprehensive policy evaluation scheme for integrated networks characterised by a multiplicity of operators (see Table 2.1).

Table 2.1. A policy evaluation scheme for network operators

indigenous features	roles of actors		barriers		synergy		
	demand	supply	demand	supply	inter-operability	inter-connectivity	inter-modality
types of modes							
product orientation							
technologies							
organisation							
regulatory regimes							
financing schemes							
pricing schemes							
product orientation							

Clearly, the implementation of this scheme would require quite some field work, in which measurable indicators would have to be gathered.

It should be added that such indicators would have to be collected over a time span which would allow for change. Thus some sort of an observatory based on a systematic monitoring of information might be needed. The overall evaluation framework might be based on four different assessment angles:
- technological harmonisation of multimodal networks,
- efficiency growth for private and public actors using these networks,
- distributional equity for all groups and regions involved,
- sustainable development in terms of environment, resources and safety.

The new challenge is thus to identify how private and public actors (and chains of actors) will use the new opportunities in the emerging European network economy, including creative ways of coping with bottlenecks.

2.5 Borders as Barriers and Gateways in Europe's Futures

In an open network economy borders should only play a modest role. In the past years many old borders have vanished and new political-economic maps have emerged. While Europe has exhibited especially fast dynamics in this respect, other continents (e.g., NAFTA in North-America) are gradually following the same trend. This means that the ongoing process of economic integration and economic competition in an open network economy is creating new roles and new possibilities for national states, cities and regions. Barriers related to former borders may disappear, but national self-interest may create new barriers. Thus establishment and renewal are co-existing with one another (see Nijkamp 1994).

Policy-makers find themselves in a difficult position as the deregulation paradigm may prevent them from direct intervention. Controllability via public agencies becomes more and more problematic. Cities and regions tend to form their own strategic alliances without much consideration for the former borders of nation states. However, transborder co-operation may generate unexpected benefits, as the economies of scale of new strategic alliances across the borders are significant (see Ratti and Reichman 1993). Consequently, borders in a permeable network are not necessarily barriers to development, but also windows of opportunities. This applies not only to commercial activities, but also to the exchange of information and knowledge. It remains generally true that borders and barriers lead to a lower performance of a network. Borders exist for geopolitical reasons and barriers exist because of institutional, physical or human-made impediments. They form an obstacle in a free transfer of people, goods or information. Some of these impediments are given by nature (mountains, lakes) of course, but most of them are man-made and created for the sake of convenience or protection or are unintended effects or spin-offs of other barriers. Examples of man-made barriers are: congestion, fiscal constraints, institutional rules, technical conditions, market regulations, cultural inertia, language barriers or information shortages. All such barriers hamper competition in an open network economy.

In Europe, the traditional patterns of competition – within national borders – are being replaced by vigorous competition on a multi-national scale, as traditional frontiers disappear. Regions of different countries are becoming part of a *transnational economic network*. These developments may lead to a tendency in which established economic centres are losing part of their innovative potential in favour of regions with *medium-sized cities*. For example, the network economies in the French regions Provence-Alpes-Côte-d'Azur and Languedoc-Rousillon are based on innovative small and medium-sized companies which maintain linkages among themselves and with large enterprises. Here various forms of expertise in *collaborative networks* transcend the older types of industrial strategy based on *internal concentration*. Besides these French regions the "Third Italy" is an example of a territorial network of small business maintaining more or less formal relations.

The geopolitical changes at the regional level do not only concern the position of European centres (e.g., the shift from Bonn to Berlin, or the emergence of new capital cities in the former Yugoslavia and USSR), but also the former border areas. The internal border areas in the EU are likely to receive sudden improvements in their competitive positions because of shifts from geographical 'dead ends' to new gateways. The external border areas do not have such perspectives, so that their peripheral position may even be aggravated as a result of more integration and cohesion inside the EU, unless new transnational networks to the East are built. Furthermore rural areas, coastal areas and islands will be facing many new challenges with a clear goal of achieving a better structural position in the 'Europe of Regions' (see EC 1991, Amsterdams Historisch Genootschap 1992).

Economic history shows that Schumpeterian waves of economic restructuring appear to discriminate among various regions or cities. In the past decade especially the *information and communication sector* has often been regarded as the key sector in the so-called fifth Kondratieff wave. The knowledge and information component appears to be extremely important in this new technology sector, and this has led some authors to the conviction, that so-called 3C-plus regions (regions with creativity, competence and connectivity) are the most promising areas for spatial-economic dynamics. On the other hand, the losers in this game will be the 3C-minus regions which are characterised by congestion, criminality and closure.

The creation of a *'new technology'* niche in a region is often regarded as a guarantee of regional revitalisation. However, the regional innovation potential is a multi-faceted phenomenon which shows much variation, as is witnessed by the Silicon Valley, the Greater Boston area, the London-Bristol corridor, the Dutch Randstad, and the greater Barcelona area.

The expected changes in the European scene have generated a broad interest in the future economic maps of Europe, based on 'plausible rational speculation'. Examples of such new maps are the 'blue banana', the 'blue star', the 'green grape', the 'Euregg-model' and many others (an overview of various European maps is contained in Nijkamp et al. 1993). Such maps are not meant to be blueprints or predictions, but thought experiments based on plausible scenarios (economic, social, political, technological etc.). Various interesting sketches of European spatial developments can be found in Brunet (1991). In Figures 2.3 (a)–(c) three illustrations of such scenario thinking are given.

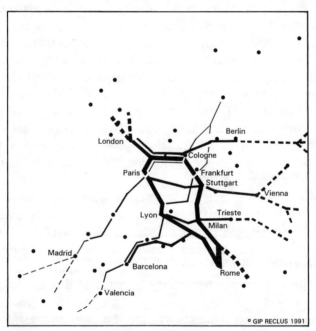

Figure 2.3(a). A European 'business as usual' scenario

The European Network Economy   37

Figure 2.3(b). A European despair scenario

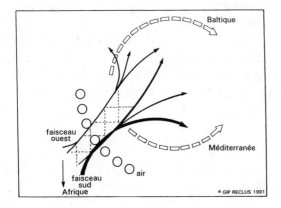

Figure 2.3(c). A European continental scenario

The first scenario is a 'business as usual' scenario where existing force fields based on economic efficiency are reinforced ('la poursuite des tendences naturelles'). A 'despair scenario' is depicted with the Mediterranean edge economically cut off from the rest of Europe and the force field oriented to the North-West ('le scénario du désespoir'). A compromise is sought in a continental integration scenario stretching from North-Africa and the Mediterranean to the Baltic area ('modéle strategique des liaisons éuropéennes').

The main intention of such scenario maps is to identify which policy actions (spatial, economic, technological) are necessary to cope with the negative aspects of such scenarios were they to take place. A common set of policy strategies which would be necessary in all cases is difficult to identify. It is plausible that a European network policy would include European dimensions in local and regional policy-making as well as free access to multi-modal European networks in order to stimulate international competition rather than to protect local and regional home markets. Technological standardisation within and between all infrastructure networks in all European countries should be promoted in order to reap the fruits of real international integration in Europe. The policy would call for the removal of all bottlenecks (institutional, technical) which hamper a real interoperability of intermodal network configurations in Europe with due consideration to all negative externalities of a mobile Europe. For instance, this could be achieved by favouring a more efficient use of existing transport networks (e.g., via telematics) to the detriment of an expansion of physical capacity. The network policy would ultimately aim at creating a balanced development between small and medium localities and large internationally-oriented metropolitan areas. Only a European network based on 'glocalisation' will be able to generate benefits which stem from the specific socio-political and economic geography of Europe.

References

Amsterdams Historisch Genootschap (1992), *The United States of Europe*, Amsterdam.
Brunet, R. (1991), *Vers les Réseaux Transeuropéens*, RECLUS, Paris.
Capello, R. (1993), *Spatial Aspects of Telecommunication Network Externalities and Regional Development*, Avebury, Aldershot.
Capineri, C. (1993), 'Italian Geography and Networks: A Bibliographical Essay', *Rivista Geografica Italiana*, 11:1, 5–15.
EC (1991), *Europe 2000*, Brussels.

EC (1992), Master Plan for the Road Network and Road Traffic, Brussels.
Hall, D. (1993), 'Impacts of Economic and Political Transition on the Transport Geography of Central and Eastern Europe', *Journal of Transport Geography*, 1:1, 20-35.
Kamann, D.J., and P. Nijkamp (1991), 'Technogenesis', *Technological Forecasting and Social Change*, 39, 35-46.
Knowles, R. (1993), 'Research Agendas in Transport Geography for the 1990s', *Journal of Transport Geography*, 1:1, 3-11.
Masser, I., M. Wegener and O. Svidén (1992), *The Geography of Europe's Futures*, Belhaven, London.
Nijkamp, P. (1993a), 'Towards a Network of Regions: The United States of Europe', *European Planning Studies*, 1:2, 149-168.
Nijkamp, P. (ed.) (1993b), *Europe on the Move*, Avebury, Aldershot.
Nijkamp, P., G. Pepping, and A. Reggiani (1993), 'Examples of European Transport Networks and Corridors', *Technology Reviews*.
Nijkamp, P. (ed.) (1994), *New Borders and Old Barriers in Spatial Development*, Gordon & Breach, London.
Nijkamp, P., J. Vleugel, R. Maggi and I. Masser (1994), *Missing Transport Networks in Europe*, Avebury, Aldershot.
Ratti, R. and S. Reichman (1993), 'Spatial Effects of Borders', in P. Nijkamp (ed.), *Europe on the Move*, Avebury, Aldershot, UK, 68-87.
OECD (1992), *Advanced Logistics and Road Freight Transport*, Paris.

CHAPTER 3

Regional Development, Federalism and Interregional Co-operation

Riccardo Cappellin
University of Rome Tor Vergata

3.1 The Case for New Regional Policies

Wide regional disparities in the economically lagging regions of the European Union persist after several decades of regional policy. This clearly indicates the need to modify the traditional policy doctrine. According to the traditional doctrine, the scarcity of capital and qualified labour as well as barriers associated with economies of scale cause the bottlenecks to development. Modern approaches to regional development, such as the endogenous development approach and the network approach, suggest a different diagnosis. The most important obstacles to development are threefold: first, low entrepreneurship and innovation capabilities, secondly scarcity of a specific know-how and thirdly, the low internationalisation of less developed regional economies.

According to the traditional doctrine, regional policies should defend the weakest regions from international competition. Their main instruments are financial transfers from the most developed regions in order to reduce the costs of the factors of production and to sustain the price competition of firms in less developed regions. Contrary to this traditional doctrine, regional policies based on the "endogenous development" and the "network" paradigms take a more active strategy. Their aim is to reduce the technological and organisational gaps between the weakest and the strongest regions. These frameworks emphasise the concept of dynamic competition based on the capability to adopt innovations and to promote the internationalisation of firms. Also, they imply the need to reduce the various adjustment and transaction costs which cause barriers to entry and growth in firms located in less developed regions (Cappellin 1991; Cappellin and Tosi 1993). High adjustment costs imply that the speed in adopting innovation is too low, and therefore the gap with more developed regions cannot be reduced. High transaction costs hinder various forms of agreements with external firms and organisations. In addition, appropriate transfers of know-how and financial resources from outside are impeded.

The endogenous and network approaches target policies which promote measures related to service activities, while traditional policies have concentrated only on industrial activities. This orientation is due to the fact that modern producer services, such as R&D, consulting services and also financial and marketing services contribute to those organisational and social changes which are required in the adoption of innovations. They also facilitate the flows of goods, people, capital and information between a region and the outside world. Furthermore, modern approaches to regional policy emphasise up-to-date infrastructures, which is a key prerequisite for the conversion from traditional to more modern productions and in the integration of isolated regions into the international economy. Moreover, as services and infrastructures are typically concentrated in urban centres, regional development policies based on the endogenous development and network approaches imply a greater degree of integration between economic and territorial policies.

Regional policies based on the decrease of adjustment costs and transaction costs have two main aims: they focus on influencing both the general environment in which firms operate and on improving their capability to make timely and adequate decisions. Thus, modern approaches to regional policy do not have a once-for-all effect on costs to be rapidly compensated by the increase in factor prices, but they have a long term effect in allowing a stable increase of the growth rate in productivity, production, and sales. These new regional policy strategies may be compatible with (or even complementary to) competition policy because they aim at equalising the opportunities for entrepreneurial behaviours in different regional environments.

On the other hand, the competition policy of the European Union may represent a useful constraint in reorienting the strategies of regional policy. The development of the weakest regions in the EU requires an effort to reintroduce market factors in these economies in which, similar to former socialist countries, a pervasive system of state intervention has almost completely distorted the price structure and the motivations of microeconomic behaviour. Thus, the EU competition policy may have some positive effects in limiting measures, which do not have any long-term effect on economic development but are only a form of public assistance and instrument for the redistribution of income.

The challenge of the European Union's competition policy is to contribute a new strategy for regional policy, which may be defined as a "market oriented regional policy" based on the criteria of economic efficiency. This kind of regional policy should aim to support the capabilities of local, existing or new, firms to sustain international competition by continuously facing internal and external obstacles. It should not attempt to protect regional firms through financial subsidies which creates a permanent obstacle to competition.

Overall, modern regional policies are more complex than simple financial transfers, and they require higher technical and organisational capabilities in the EU as well as in national and regional administration. In fact, the backwardness of the economic system of the weakest European regions seems also to imply the technical, organisational, economic and cultural backwardness of the regional policies which have traditionally been implemented in these regions. Given this dilemma, the shift towards a new strategy of regional policy seems to be insufficient unless there is a parallel shift towards new institutionalised structures of relationships between regions and nation states. In particular, the national level is clearly inadequate for organising regional policies which require a tight contact with the different regional realities. Therefore not only the attempt to pursue the traditional policies, but also the attempt to change regional policy strategies while maintaining the traditional centralised structure, will lead to poor performance.

The removal of national barriers within the European Union and the development of economic and political relationships with the European countries external to the EU implies a new geo-economic order in Europe. In addition, it contributes to a change in hierarchical relationships among various regional and urban production systems (CEC 1991; Cappellin and Batey 1993) as well as between regional and national and EU institutions.

3.2 The Europe of Regions

In the conditions of the new geo-economic order, global competition does not only concern individual firms but also various national and regional production systems. In this perspective, there are emerging groups of regions and countries at the transnational scale which are characterised by a higher than average level of economic integration. These large European transnational meso-regions can rely on historical traditions and values, which can define a common identity and a sense of belonging. They can foresee a specific common role in the large European economic space, where there exist different relationships at different levels. Co-operative relationships based on specialisation and complementarity of each individual region or urban centre prevail within each regional and urban network. Consequently, at the European level, competitive behaviour prevails in the relationships between individual urban networks (Cappellin 1988a, 1989 and 1991a).

There is an increasing consensus that basing European integration on the two traditional conceptions is too restrictive. Indeed, the transnational interregional perspective challenges the conception of the Single European Market (or a European Economic Space) and a European Community.

The Single European Market is characterised by well-known features: the abolishment of obstacles to the international movements of goods, services, capital and people, as well as the creation of a common legal framework and a monetary system. These are supposed to lead to a convergence of inflation rates and other macroeconomic indicators, which would ensure a stable framework for exports and international investments. The traditional conception of a European Community with regard to the political and institutional development of European integration has been elaborated since the 1950s (Cappellin 1988b, 1990, and 1991c). It differs from the Single European Market or a European Economic Space by devoting more attention to microeconomic aspects and the belief that specific public interventions are needed in the process of European integration. This conception is conditioned by several factors. Overcoming the national conflicts which had led to two world wars played a primary role in its development. The harmonisation of national regulations and substituting a European legal system was a logical consequence. The conception of a European Community has also relied on the merging of local cultures and the overcoming of national barriers, and thus contributed to developing a European cultural identity and a European nationalism. In addition, it has put emphasis on the development of supranational institutions with hierarchical power with respect to individual nations. This has also implied the gradual extension of the powers of the European Community in those fields which are the competence of regional institutions.

Whereas traditional conceptions of the European integration increasingly seem to conflict with the actual trends in the European economy and society, a new conception has gained strength since the 1980s. The *Europe of Regions* or the *European Federalism* has gained impetus from the model of flexible production and the development of network organisational forms of business and government not only at the local level but also at the interregional level. It has also put emphasis on the claim of autonomy and self-government made by regional institutions and on the increasing consensus of the principle of subsidiarity. This derives from the perception that regional and national diversity is valuable as well as the development of transnational historical, social and cultural identities. Another characteristic feature of the federalist conception of European integration is a flexible geometry of interregional and international co-operative relationships. The development of non-hierarchical relationships based on the partnership between national and regional institutions in joint programmes, as well as the development of specific European programmes are also characteristic to the Europe of Regions.

Clearly, the typically federalist conception of the Europe of Regions differs from that of the European Single Market. The difference is due to the fact that interregional flows are neither conceived in a functional way nor are they controlled by a global system of multinational or transnational firms. Rather, they are seen as the effects of interdependence and integration among the different regional productive systems. Compared with a supraregional European Community, according to the "Europe of Regions" conception regionalism and claims for regional self-government are based on the belief that economic development in individual areas mostly depend upon the capability of local initiative to exploit synergies among the local resources.

The process of interregional co-operation is based on a grassroots approach and has profound implications for the characteristics of European integration. This establishes the need for a thorough analysis of concepts such as regional administrative autonomy, regionalism, federalism and nationalism, which are often used in a contradictory manner. The analysis is summarised in Table 3.1.

Table 3.1. Organisational forms and models of institutional integration

	FREE MARKET MODEL	CENTRALISTIC MODEL	FEDERALIST MODEL
ORGANISATION FORMS	market	hierarchy	co-operation
	atomistic competition	mass production	flexible production
ORGANISATION PRINCIPLES	initiative	authority	self-government
	responsibility	legal rights	synergy
	efficiency	economics of scale	flexibility
INTERACTION LOGICS	competitiveness	homogeneity	differentiation
	monetary exchange	control and dependence	influence and leadership
	interdependence	co-ordination	negotiation
GEOGRAPHICAL FRAMEWORK	homogenous space	administrative units	territorial systems
NEGATIVE EFFECTS	egoism	bureaucracy	conflicts
	economicism	assistance	assemblearism
	liberism	dirigism	veto power
POLITICAL IDEALS	liberte'	egalite'	fraternite'
INTERNATIONAL RELATIONS	free trade	mercantilism	complementarity
EUROPEAN INTEGRATION	single market	supranational community	interregional federation
NEGATIVE DEVELOPMENTS	economic and political disequilibria	nationalism and separatism	confusion and impotence

In abstract terms, the distinction between the three different concepts of the European Single Market, European Community, and Europe of Regions resembles that of the three different organisational forms in the modern theory of the firm. There is also a similarity with the three principles at the base of liberal-democratic thinking. Therefore, it is possible to identify three different institutional models or paradigms. These are each characterised by an internal logic, implying a tight interaction between concepts associated with relationships among firms and forms of political and institutional relationships. In particular, it is worth underlining the close relationship between federalism and modern organisational forms of the economic system and the individual firm. Federalism and regionalism correspond to a paradigmatic transformation of the structures of society, the economy, and production technologies in Europe (Cappellin 1990 and 1991b). Federalism ensures a larger decentralisation in the decision making process. Therefore, it represents an institutional form suitable for a socio-economic system which is better articulated, culturally more advanced and technologically more complex. While federalism is typical of an open system, centralism is typical of a closed system.

The internationalisation of economies and the integration of local and national production systems complements regionalism and federalism in the governance structures of the economy. In fact, it is typical of regional economies to imply a close integration not only of product markets, but also the flows of the factors of production. Thus, the internationalisation process transforms nation states into large regions which are internally highly heterogeneous. This may require the regionalisation of individual national systems.

It is important to stress the complementarity between the concepts of co-operation and solidarity with those of regionalism and federalism. Co-operation and solidarity can be based upon a consensus between individuals. Thus, there is no need for a superior authority, relying on legal rights to public transfers, to determine equity. Differing from widespread opinion, separatism is more connected to centralism and nationalism than to regionalism. Furthermore, centralism opposes those flexible forms of integration which are a characteristic of regionalism. It also implies the oppression of ethnic minorities which may lead to separatism, and is, in fact, a form of micro-nationalism. Thus the centralised power of the nation state is not only the main factor for disunity at the European level, but it is frequently a decisive factor leading to the division of individual national communities.

3.3 The Principle of Subsidiarity

The economic foundation of federalism is represented by the principle of subsidiarity. According to this principle, each function should be attributed to the lowest efficient decision-making level within the hierarchical system of relationships between regions, nation states and the European Union. Therefore, functions should not be transferred to a superior level when they can be efficiently exercised at a lower level. The subsidiarity principle implies a limitation of powers of national governments and the European Union. Being freed from disparate competencies, there can be gains in terms of greater flexibility, and they may concentrate their efforts in policy fields which have specific national or European dimensions. In a modern national economy in Europe, certain policy decisions are best made at the regional level. These are: firstly, territorial planning, infrastructure and environment; secondly, vocational and superior education and applied research; and thirdly, industrial and innovation policy for SMEs. Unfortunately, the inefficiency of national administrations in these policy fields is clearly demonstrated in most countries.

In addition to its relative ambiguity, a further limitation of the subsidiarity principle is its hierarchical character. It explicitly takes into account only vertical relations between the regional and the national levels and the EU level. Therefore, a strict adherence to the subsidiarity principle would imply that problems which have a superregional dimension would be dealt with at superior levels. This is a serious limitation, as most problems clearly have interregional spillover effects across regional boundaries (Cappellin and Batey 1993). However, this limitation can be circumvented: interregional co-operation both in a bilateral and in a multiregional framework offers an institutional and organisational solution. In addition, it is more efficient than the creation of new authorities to tackle policy interventions which, although having a superregional dimension, do not have a clear national relevance. Another kind of problem arises in cases in which only a limited number of regions of a given country have a common interest in the problem being considered. In dealing with relationships between border regions, co-ordination of national administrations would often produce additional problems in contrast to direct negotiation between the neighbouring regional governments. This is due to the fact that national governments know much less about the concrete problems at hand.

The problem of dividing decision-making between different levels is illustrated in Figure 3.1, in which the various functions are organised according to a hierarchical principle. Starting with functions which imply a smaller geographical unit for planning, it proceeds towards those which use a larger territorial framework for their efficient management. The horizontal axis indicates various location points, which correspond to different regional administrations.

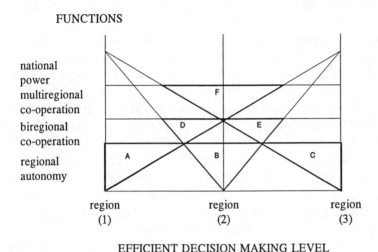

Figure 3.1. Functions and efficient decision making levels

First, the relevant areas do not intersect for some functions, and total autonomy can be allowed to each regional administration (A, B, and C). Secondly, in the case of superregional problems, i.e., when the required minimum planning units overlap, bilateral or multilateral co-operation schemes may be a more efficient solution than the delegation of all the power to a superregional or national "Authority". This is indicated by the areas D, E, and F in Figure 3.1. Thirdly, power can be delegated to a superregional or national "Authority", which may have its own legitimacy and act autonomously from the various regional governments. However, this should be done only when the area of overlap among the regions considered represents a large enough portion of their territory. In general, the principle of interregional co-operation is coherent with a bottom-up decision making process. It is the logical extension of the principle of subsidiarity. In fact, the impetus for co-operation comes from individual local government units and also from individual firms and local lobbying groups. Interregional co-operation is both an instrument and an effect aiming to promote local activity.

3.4 Benefits and Costs of Interregional Transnational Co-operation

The relationships among border regions have been and still often are relationships of conflict due to the existence of ethnic minorities, fear of immigration, fear of unfair competition, and negative environmental spillover effects. However, these relationships may become co-operative, whenever bilateral relationships are interpreted in a European context (Maillat 1990; Cappellin 1991b; Cappellin and Batey 1993; Nijkamp 1993). The design of a new interregional framework in a European perspective leads to new types of relationships between nation states and regional governments. While this is especially important for land-border regions, which are contiguous, it also applies for sea-border regions and more generally for those regions which may be considered members of a large transnational European meso-region.

In the present context, it should be emphasised that there are both benefits and costs to the process of transnational co-operation between border regions. For instance, the preservation of national barriers may in some cases represent an advantage not only for the capital regions where the national administrations are concentrated but also for some border regions. This results from their barrier position, as they are the location of custom administration, various international transport and financial activities, and military installations, and consequently may receive considerable subsidies from national governments. Notwithstanding these benefits, border regions are generally penalised by the existence of various obstacles which may be defined as territorial costs arising from "Non Europe", according to the terminology used in the construction of the European Single Market. Thus, they deserve more attention by the EU and national governments aiming to promote a greater cohesion of the European economy. This also explains why the European Commission has recently supported the development of co-

Regional Development, Federalism 51

operation schemes among regions through the INTERREG programme and the Regions and Cities for Europe (RECITE) initiative.

Frontiers, including sea-frontiers, between countries and regions should not be barriers or lines of potential conflicts, but interfaces and gateways between various European countries and also between these and the non-European countries of the Mediterranean basin. Border regions may become important hubs or nodes in the European transportation and communication networks. Moreover, border regions, for their strategic positions, may become European transnational meso-regions. These may favour the process of integration and help various individual regions to have a common role to play in the European economic space. In sum, co-operation between border regions may be consistent with the achievement of the three different objectives. First, it removes barriers between the regions considered, and consequently, solves bilateral problems and conflicts. Secondly, border regions may become an interface area or a gateway in the relationships between the two countries. Thirdly, these regions may represent an overall transnational region, which may play a role in international competition with other European meso-regions.

In general, the organisation of external relations in border regions with respect to the national community and to the neighbouring foreign regions may be specified according to two solutions - as in the organisation of external relations of a firm. The first solution, which is similar to the hierarchy solution, or the vertical integration of a firm, is the national unity solution. The opposite solution, which resembles the market solution or the monetary exchange aspect for a firm, is the regional integration, or the creation of a transnational region together with the neighbouring foreign regions. This trade-off is illustrated in Figure 3.2.

The actual solution is a compromise between the two extreme alternatives, regional autonomy and national unity. It may be defined as the search for the optimal level of regional autonomy ($0 < A < 1$), when both the benefits (B_r) and the costs (C_r) of regional integration, and the benefits (B_n) and costs (C_n) of national unity vary with the level of regional autonomy (A) as indicated in Figure 3.2. The benefits of regional integration (B_r) may result from communalities and complementarities among regions. Additional benefits include the establishment of network economies for the circulation of information and know-how, economies of scale in public services and infrastructures, and co-ordinated actions to protect joint interests. The costs (C_r) may arise from transaction costs in interregional negotiations, displacement of firms and production to neighbouring regions, and immigration from neighbouring regions. These are compared with benefits and costs of a national unity. The benefits (B_n) are derived from several factors such as protection from foreign threats by national institutions, location of custom functions in border areas, and financial transfers from national institutions. The respective costs (C_n) are due to effects of international conflicts in border areas, transaction costs in co-ordination with national institutions, and discriminatory national policies vis-a-vis peripheral regions.

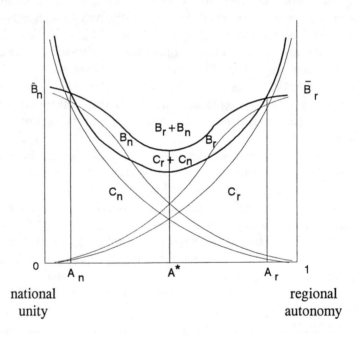

Figure 3.2. The trade-off between national unity and regional integration

When realistic assumptions are made regarding these benefit and cost functions, as those indicated in Figure 3.2, the optimal level of autonomy (A^*) is in the interval $A_n A_r$. This suggests a wide range of compromise solutions, which are superior to extreme forms of national unity or regional autonomy.

3.5   The Organisation of Interregional Programmes

The development of a strategy of interregional co-operation is hindered by a series of obstacles which in some cases represent constraints which are hard to overcome and in other cases represent preconditions for strategies of interregional co-operation. The obstacles, which have generally been identified through previous cases of interregional co-operation include differences in technological or development levels, institutional competencies and financial resources. Languages and cultures may present problems as well. In addition, reciprocal trust, competence concerning the process, and institutions to manage the co-operation scheme may be weak and therefore hinder co-operation. The urban structure of a region is also a potentially important factor. It may be open to the international economy, and municipal authorities may take a very active role in promoting interregional co-operation initiatives.

The creation of a co-operation agreement is not a sudden occurrence but a gradual learning process. Thus, the factors leading to the continuity and success of an agreement or to its interruption may be more important than the results of the agreement itself. The existence of different competencies does not represent a crucial obstacle for the creation of successful co-operation schemes. What clearly matters is the behavioural logics of the various regional administrations, as interregional co-operation requires a pragmatic logic of public entrepreneurship or project design capability, rather than the formal respect of bureaucratic competencies.

Interregional co-operation has profound implications both for policy design and for policy implementation by local authorities. It is an instrument capable of identifying innovative projects and mobilising an enterprising spirit both external and internal to the region. In particular, there is a tight relationship between the capabilities of networking at the interregional scale and those of networking at the regional level. In fact, the development of interregional co-operative relationships clearly implies the capability by the individual regional governments to establish co-operative relationships with other regional institutions and organisations. These may include partners from professional associations, individual private firms, as well as cultural and research institutions in order to design joint projects which may promote the international role of the region concerned.

Therefore, an integrated local production and innovation system represents both a strength and a prerequisite in order to participate to more complex initiatives of networking at the interregional level. The most efficient and long lasting interregional co-operation schemes are those which have an intersectoral character, and are not focused on a narrowly defined objective. They are capable of mobilising the largest part of human, productive and organisational resources which exist in a region (Cappellin 1991b). Overall, interregional co-operation may stimulate countries in which economic policy attends only to sectoral or macroeconomic issues to recognise the importance of regional policies. In other countries interregional co-operation may stimulate change in outdated regional approaches, such as those still followed in many economically lagging regions in Europe.

It is important to note that interregional networking and co-operation have implicit hierarchical characters. They connote the existence of regions performing the role of leaders especially in preparatory phases and the creation of specific structures or contractual provisions which will ensure co-ordination in implementation. In the case of economically lagging regions of the EU, interregional co-operation initiatives are hindered by lack of a project-oriented mentality, which has resulted from the prevailing centralistic and dependent development model. In these regions it is often still not perceived that interregional co-operation may be a business also for private organisations. Local actors only aim at the financial incentives to be granted by public authorities within the framework of these initiatives, and refuse to take a more active role in the organisation of co-operation projects. Thus it is necessary that co-operation initiatives be promoted primarily by the various regional administrations with the technical assistance and the financial help of national and EU institutions.

Essentially, the development of a strategy of interregional co-operation would help overcome a centre-periphery logic, which is prevailing in economic lagging regions. Primarily, it would stimulate these regions to adopt a polycentric logic. As a result, regions which are peripheral with respect to a supposed centre of the European economy may also become central in a different geographical perspective. This perspective may be enlarged to also include countries which do not belong to the European Union, such as those of Eastern Europe and the Mediterranean basin.

The centre-periphery logic is negative in the sense that it leads to efforts which focus on the development of further centripetal relations between each peripheral region and a supposed centre. Instead, horizontal relations of interregional co-operation should be developed. This would undermine the present practice in which economically lagging EU regions and also eastern European countries compete with each other in order to attract resources and attain the best positions in the process of economic integration with the central countries and regions in the European economy.

It should be realised, however, that in some cases the logic of centre-periphery may also represent the stimulus for some regions to create an alliance with a European mesoregion for purposes of avoiding marginalisation within the new European economy. Regions and countries which are promoting co-operation schemes within the Atlantic Arc or the Baltic area are illustrative examples of this phenomenon.

In general, the logic of interregional co-operation has a selective or hierarchical character, as the regions which are most capable to participate in a networking process at the European level are the most developed ones. This may lead to an increase in regional disparities, unless national and EU institutions help economically lagging regions to escape from their relative closure and increase their awareness of the needs of each region to develop its international relations.

3.6 Conclusions

The creation of the European Single Market and the new relationships with East European countries oblige the various European regions to verify their capabilities to exploit new opportunities and to tackle new challenges. The paradigm of territorial networks emphasises the need to include the effects of new trends, such as the process of internationalisation of regional economies, into the analysis of regional development. A similar new trend is the diversification of the European economic space into different subsystems or meso-regions made by various regions and urban centres. In this context, the gradual development of schemes of interregional co-operation is a powerful instrument in order to promote European integration. This can be seen in the case of central and developed regions and in that of peripheral and economic lagging regions, both inside the EU and between EU regions and their external neighbours.

While national authorities have a vested interest in the preservation of international differences, the creation of new large transnational regions based on interregional co-operation may promote the overcoming of national barriers. National and EU institutions should have clear interests in promoting cross-border co-operation in order to promote a more integrated European territory. In addition, with regard to equity, interregional co-operation represents an effective strategy for promoting regional development in peripheral regions. Assigning a more vital role to interregional co-operation may contribute to the adoption of new regional development strategies in economically lagging regions. These strategies are based on the endogenous approach, the responsibility of local actors and the development of internationally competitive activities. It should be underlined here that regional problems have not only an economic dimension but also an institutional dimension. Regional governments advocate greater powers in the management of their respective economies from national and EU institutions. The belief in the ideals of autonomy, regionalism and federalism is increasing among the European peoples.

Interregional co-operation especially with regions of former socialist countries is instrumental in strengthening democratic institutions and thus reinforcing political stability in Europe. In addition, interregional co-operation may also be of fundamental importance in reinforcing the powers of local governments, and their trust in their own capabilities. It may also support the sense of responsibility, the values of democratic accountability, and the efficiency of local public administrations in many less developed regions of Western Europe.

The network approach in regional policy indicates the usefulness of promoting co-operation among the different EU, national and local institutions in the creation of common infrastructures and services and in the implementation of various common programmes. These inevitably have an impact on all three vertical levels (cf. Figure 3.1). Therefore, EU regional policies and in particular, the regional development policies of economically lagging regions within the EU should offer effective partnerships between regional actors. In addition, they must be compatible with the preservation of regional differences, which are themselves one of the main resources of the European Union.

References

Cappellin, R. (1988a), 'Transaction costs and urban agglomeration', *Revue à Economie Regionale et Urbaine*, 2.
Cappellin, R. and W. Molle (1988b), 'The coordination problem in theory and policy', In Molle, W. and R. Cappellin (eds.), *Regional impact of Community policies in Europe*, Avebury-Gower, Aldershot.
Cappellin, R. (1989), 'International linkages among cities: a network approach', Paper presented at the 29th European Congress of the RSA, Cambridge, 1989, Published in Italian as 'Networks nelle città e networks tra le città', In Curti, F. and L. Diappi (eds.), *Gerarchie e Reti di Città: Tendenze e Politiche*, Franco Angeli, Milano.
Cappellin, R. (1990), 'The European Internal Market and the Internationalisation of Small and Medium Size Enterprises', *Built Environment*, 16, 1, 69-84.
Cappellin, R. (1991a), 'International networks of cities', In Camagni, R. (ed.), *Innovation networks: a spatial perspective*, Belhaven, London.
Cappellin, R. (1991b), 'Theories of local endogenous development and international co-operation', In Tykkyläinen, M. (ed.), *Interregional co-operation and development in the European fringe areas*, Gower, Aldershot.
Cappellin, R. (1991c), 'Patterns and policies of regional economic development and the cohesion among the regions of the European Community', In Leonardi, R., Report to the European Commission on *The State of Social and Economic Cohesion in the Community prior to the Creation of the Single Market: the View from the Bottom-Up*, Bruxelles.
Cappellin, R. and Batey, P. (eds.) (1993), *Regional Networks, Border Regions and European Integration*, Pion, London.
Cappellin, R. and Nijkamp, P. (eds.) (1990), *The spatial context of technological change*, Avebury-Gower, Aldershot.
Cappellin, R., and A. Tosi (eds.) (1993), *Politiche innovative nel Mezzoqiorno e parch tecnologici*, Franco Angeli, Milano.
Commission of the European Communities (1991), *Europe 200: Outlook for the development of Community's territory*, Bruxelles.
Maillat, D. (1990), 'Transborder regions between members of the EC and non-member countries', *Built Environment*, 16, 1, 38-51.
Nijkamp, P. (1993), 'Border regions and infrastructure networks in the European integration process', *Environment and Planning C*, 11, 431-446.

CHAPTER 4

Remaking Scale: Competition and Cooperation in Prenational and Postnational Europe

Neil Smith
Rutgers University

4.1 Introduction

The geography of Europe only two decades ago was broadly conceived as a stable hierarchy of places at different spatial scales: Eastern and Western blocs, discrete nations, subnational regions, and local and urban communities. The disruption of this "given" postwar geography in the intervening two decades and of the political, economic and cultural assumptions that went with it could barely have been predicted in the early 1970s (but see Mandel 1975, 310–42 for a prescient discussion; Rowthorn 1971; Murray 1971). Certainly the development of a "European Economic Community", equalizing conditions of trade in several commodities between six countries beginning in the early 1950s, and the steady growth of a more fully fledged "Common Market" served notice that some disruption of the traditional economic geography (at least at the national scale) was afoot. Nonetheless, the reconstruction of Europe at all spatial scales that would follow the 1970s economic depressions in the West and the post-1989 implosion of official Communist Party rule in the East were quite unforeseeable. Thereby, the largely economic evolution of the Common Market into the European Community in the 1970s and 1980s and now into the more politically inspired European Union was bound up with a much more complex and halting entanglement of social, cultural and political as well as economic restructurings.

In the mountains of commentary about a "New Europe" – some optimistic to the point of fantasy and some so pessimistic that they depict a negative fantasy of apocalypticism – the shifting geography of Europe and especially the radical reorganization of the geographical scale of various kinds of societal activity has either been neglected or treated as a series of disparate but hardly connected events. The restructuring of geographical scale, however, has been at the centre of the political, economic and social redefinition of the "New Europe". The politics of scale, it turns out, helps to provide not only a theoretical perspective on the hotly contested questions of competition and cooperation but gives a certain fix on the direction of change in a remade Europe.

4.2   Theories of Geographical Scale

In order to deal more effectively with the implications of a remade Europe from the vantage point of scale, it is necessary to consider existing theories of scale – or more properly, the social production of scale. Geographical scale is traditionally treated as a neutral metric of physical space: specific scales of social activity are assumed to be largely given as in the distinction between urban, regional, national and global events and processes; and analysts choose specific scales as appropriate for examining specific questions. There is now, however, a considerable literature arguing that the geographical scales of human activity are not neutral "givens", not fixed universals of social experience, nor are they an arbitrary methodological or conceptual choice (Taylor 1981; Smith 1984; 1992; Marston 1990; Paasi 1991; Herod 1991; 1992; Jonas 1994). Rather, scale should be seen as materially real frames of social action. As such, geographical scales are historically mutable and are the products of social activity, broadly speaking.

To illustrate this simply, it is only necessary to compare the classic European walled city of the medieval period with the contemporary conurbation of Los Angeles or Sao Paulo. Both the walled and the contemporary city symbolize what we take to be the urban, but the scale of urban life in walled London is radically different from that of Sao Paulo today. This leads us to a refinement of the axiom that scale is a materially real frame of social action: geographical scale is socially produced as simultaneously a platform and container of certain kinds of social activity. Far from neutral and fixed, therefore, geographical scales are the product of economic, political and social activities and relationships; as such they are as changeable as those relationships themselves. At the very least, different kinds of society produce different kinds of geographical scale for containing and enabling particular forms of social interaction. The medieval city is the locus of feudal commerce and simultaneously a place to be defended from external military attack, while the modern metropolis is much more the expression of an expansive capitalism premised on large scale production, widespread financial, service and communication networks, and mass consumption. Scale is the geographical organizer and expression of collective social action.

If this is a reasonable way of rendering scale a simultaneously historical and material artefact, the next set of questions presumably focuses around the ways in which scales are actually set or fixed amidst the flux of social interaction. Here I think geographical scale is best conceptualized as the *spatial* resolution of contradictory social forces; in particular the resolution between opposing forces of competition and cooperation (Smith 1984). Take, for example, the nation-state. The boundaries of the nation-state represent a *geographical* bounding between those places and actors who are prepared to cooperate vis-a-vis certain social requirements and those with whom competition is the determining relationship. In the most immediate sense, most national boundaries were the product of political and/or military contest, but they were drawn precisely as a means to establish and defend territorial units of a specific economic and cultural definition. Within the nation-state, corporations cooperate broadly in the construction of governmental apparatuses determining conditions of work, legal systems, conditions of private and public property holding, infrastructure for commerce, travel to work and communications, national defense. At different scales, these same corporate entities may well compete over customers, product identity, technological advantage, markets, etc. The boundaries of nation-states became the geographical demarcation of the compromise between competition and cooperation.

Scale then can be both fluid and fixed – materially as well as conceptually. "The language of scale," Jonas suggests, "is an anticipation of the future"; "future scales eventually become the 'scale fixes' to existing and imposed scale constraints, if only to create new constraints and opportunities for domination/subordination" (Jonas 1994, 262). The *production of scale*, therefore, is a highly charged and political process as is the continual reproduction of scale at established levels (e.g., defense of national boundaries, community tax base, regional identity). Even more politically charged is the reproduction of scale at different levels – the restructuring of scale, the establishment of new 'scale fixes' for new concatenations of political, economic and cultural interchange. Newly fixed scales of social intercourse establish fixed geographical structures bounding political, economic and cultural activity in specific ways; highly contentious and contested social relationships become anchored if not quite in stone at least in landscapes that are, in the short run, fixed. In geography, political difference is fossilized, as it were, naturalizing whole realms of contestable social organization. "Jumping scale" – the reorganization of specific kinds of social interaction at a higher scale and therefore over a wider terrain, breaking the fixity of "given" scales – is therefore a primary avenue to power. This applies whether we are considering national claims to empire, a city's efforts to annex surrounding suburbs, or feminist efforts to dissolve the boundaries between home and community (Marston 1990; Saegert and Leavitt 1990). Thus, the demarcation of scale should be seen as absolutely central to the processes and politics of uneven geographical development.

Most significantly, it is the scale of the nation-state that is being restructured as part of the new Europe, and it is this scale that occupies us here. As Gupta has pointed out, "the nation is so deeply implicated in the texture of everyday life and so thoroughly presupposed in the academic discourses on 'culture' and 'society' that it becomes difficult to remember that it is only one, relatively recent, historically contingent form in organizing space in the world" (Gupta 1992, 63). Indeed it is only one scale at which the world is organized even now. Whereas the local scale can be conceived as expressing the geographical range of daily reproduction activities (e.g., the journey to work) and the regional scale as an increasingly relict sub-national expression of the geographical coherence (or otherwise) of distinct production systems, the scale of the nation-state springs more from the global circulation of capital. With the internationalization of commercial capital in the seventeenth and eighteenth centuries, the question of coordinating competitive and cooperative relationships between capitals became increasingly vital. The nationalization of capital, simultaneous with and as part of the internationalization of capital, was the solution that emerged historically (see Hobsbawm 1990; Nairn 1977). National capitals and their attendant political frameworks in the nation-state emerged as a vital geographical means for coordinating and arbitrating economic competition between capitals at the global scale. National capitals are in effect different national "laws of value" in a wider global market, and they remain coherent to the extent that the nation-states devised for the purpose succeed in protecting the gamut of social, economic and cultural conditions that sustain individual national capitals. That is, the functions of the state which were in earlier times attached to lower spatial scales of territorial control – city states, duchies, kingdoms, etc. – are, with the advent of capitalism, reconstituted at the scale of the nation.

Two caveats are important here. First, it would be a mistake to overgeneralize and assume a complete congruence of political and economic interests. From the American Civil War and the Paris Commune to the fate of the Hapsburg Empire and the failures of Versailles, there are abundant illustrations of the contentious fit of economic and political interests in the geographically absolute territorialization of the nation-state. These struggles took place along class and race lines or resulted from intraclass conflicts, as much as they were bound up with cultural and economic definitions of various nations in birth. Ongoing as such contests are, they did not prevent the emergence of the nation-state as the appropriate geographical scale and political means for arbitrating the division of capital.

Second, although I have stressed here the economic and political relationships that lead to the pupation of specific geographical scales, it is important to realize that the production of scale is also a cultural event. Individual and group identities are heavily tinctured by attachments to place at different scales. At the national scale, nation-states may be an expression of competing capitals in the world market, established, defended and expanded by military as well as political and economic means, but they also involve an extraordinarily deepseated identity creation (Anderson 1983). Nationalism is a cultural and ideological force in its own right which helps sculpt the spatialization of social relations from the start, and which represents at times a decisive force in any restructuring of scale.

4.3 A Post-national Europe?

It is commonly understood that the origins of the "New Europe" lie first and foremost in the altered relationship between the various national economies of Europe and the international market. At the simplest level, three interconnected kinds of shifts beginning in the postwar world but quickening in the 1970s led to the current restructuring of Europe.

First, economic globalization. Although a world market in commodity extraction and exchange was largely in place by the nineteenth century, the globalization of other aspects of the economy had to wait until quite recently. The great depression of the 1920s and 1930s, with somewhat coordinated crashes in Europe and North America and attendant effects throughout the world provoked the realization of a global financial system already partly realized. The establishment of various international financial organizations in the wake of World War II – the IMF, World Bank, GATT etc. – represented an attempt at global regulation of this emerging reality. Until the 1970s, however, no matter how global (or at least international) the commercial and financial systems had become, economic production remained largely contained within preexisting national boundaries. Only with the combined disinvestment and reinvestment that marked the global recession of the early 1970s did economic production begin in any small degree to transcend national boundaries and the national scale of political economic organization. In its most common form, the internationalization of production involves the manufacture of component parts in several different countries and their assembly at one central location. Even in the 1990s, the internationalization of production has affected only certain well publicized sectors of production – automobiles, textiles, electronic goods, for example – and even there only partially. Along with this emerging globalization of production came a dramatic increase in international labour migration and the emergence of a more accomplished international division of labour than in any previous era (Froebel et al. 1980).

The emerging globalization of production and finance as well as commerce should not be exaggerated, however. Many production activities remain local and regional or are still organized at the national level (as indeed is true with commercial and financial activity – cf. local commodity markets, local money lending, and local mortgage companies). And it is also true that amidst this internationalization a new (not just relic) level of local and regional production has at times emerged, although too much is often made of this as in the fabled case of the Third Italy (Piore and Sabel 1984). The connections between such cases of relic and new localization in production and the internationalization of the economy are complicated and cannot be investigated here. Suffice it to say that while globalization necessarily implies the corollary of localization, a clear trajectory of change toward a more accomplished internationalization (and at times globalization) of most functional sectors of the economy has emerged.

But why did such an internationalization take place? This leads us to consider the second shift that occurred in the postwar period: the increased scale of capital accumulation. Capitalist economies carry within them an inherent tendency toward an increased level of capital accumulation. This argument is probably best theorized by Marx and marxist economists at the level both of individual and collective capitals, but it is recognized equally in the neo-classical economic dictum that a healthy economy is a growing economy; a no-growth economy is an economy in crisis – in fact, a long-term impossibility given capitalist conditions of production. There is no absolute necessity for a growing national or international economy to translate into the need for an expanded scale of accumulation at the level of individual capitals, but there is a compelling logic behind such a centralization and concentration of capital (Marx 1967, chs 23–25; Smith 1984). And indeed this is what happened in the postwar world. In the European automobile industry, for example, production for individual national markets was clearly an unprofitable prospect by the 1960s, and by the 1970s a series of automobile plants were established by all the major producers with a Europe-wide or at least multinational market in view. This expansion of the scale of accumulation is precisely what drove the internationalization of production in the first place and further enhanced the internationalization of markets.

The third shift that took place beginning in the 1970s involved the unprecedented transnationalism of labour in the world economy. Not only the production functions themselves were increasingly internationalized but the labour force available for production was increasingly international. This is true not just in terms of the numerically more prevalent cases of unskilled and semi-skilled labour from outside Europe and even the peripheries of Europe, but also with more skilled labour.

Together and in myriad different ways, these three shifts have dramatically expanded the scale at which the command functions of the economy operate. As a result, the geographical congruence between economic and political functions expressed in the nation-state became more and more tenuous. Nationally established states became less useful and less convenient to internationally mobile capital; were less and less willing to expand regulations or even sustain existing regulations over capital; were less able, and despite shrill national outcries often less interested in, controlling the immigration of cheap foreign labour; and therefore found themselves more and more able to relinquish some of the national state's traditional regulatory role in social reproduction. Displaced increasingly from the national to the international scale in the world market, *economic* competition between *politically defined* and territorially fixed nation-states intensified dramatically in the 1970s. Ironically, of course, this occurred at the precise moment when the reality of "national capitals" was diffusing – precisely when the territorial definition of ruling economic and political interests was diverging. Necessity, then, was turned into virtue, and the privatization of various economies in the 1980s, as well as dramatic cuts in social services, housing and welfare provisions from Britain and the U.S. to the Netherlands and even Sweden has to be seen in this context.

If the political crises of privatization and service cuts since the 1980s are in every way a social crisis of the internationalization of capital, they should also be seen as rather desperate responses by national state governments to retain and redefine a political role for themselves at a time when their economic functions were increasingly subordinated to the global economy and even emerging aspects of a global state, especially the IMF. The power of the IMF, the World Bank and the United Nations is routinely used to discipline the states of the developing world, but in Europe the political assertion of the world market is more gingerly applied. Britain in the late 1970s felt the squeeze of the IMF as indeed does most of Eastern Europe now.

The much vaunted obsolescence of the national state therefore represents a very real trajectory of change and is of course central to the emergence of the "New Europe." In this respect, we can see clear connections between economic internationalization and the emergence of what might be called a "post-national Europe." The argument of a post-national world has been put forward more broadly by Appadurai (1993). And of course very similar forces of "scale jumping" are afoot in the tripartite ratification of NAFTA – the North American Free Trade Agreement – which will begin to provide a framework for and help routinize a pattern of transnational economic and labour movements between Mexico, the US and Canada that has been emerging at least since World War II but more intensely since the 1970s. More recently, the Southern Common Market has been ratified for South America.

4.4 The Politics of "Integration"

If the national scale of social, political and economic organization, represented a solution to a specific problem, namely how to arbitrate contradictory requirements for competition and cooperation in the economic sphere, it is important to make clear that this was not a universal problem even if it might be considered global. By this I mean that in the seventeenth to nineteenth centuries, it was specifically the commercial classes and emerging capitalist interests across the world who faced the problem of arbitrating competition and cooperation and establishing the political conditions under which capital accumulation should proceed. The organization of competing capitals was hardly an inherent problem for peasants, workers or even the petit bourgeoisie and only impinged on the aristocracies to the extent that the latter conjoined their traditionally landed wealth with commercial production capital. The emergence of the nationstate and the national scale of political and economic organization therefore represented in the first place a solution to a strictly class problem. This is obviously not to deny that the national framework of laws, infrastructure and economic development and the other accoutrement of national development, established primarily out of the political and economic aspirations of the emerging bourgeoisie, affected (sometimes but by no means always beneficially) other classes and groups as well – sometimes sooner, sometimes later. The ingenuity of the national state as a territorial form of governance is precisely that it tied clear class economic interests to limited promises of democratic participation and the dissolution of earlier forms of oppressive absolutism. Nor is it to deny that aspirations for such benefits led to the inclusion of various groups and classes in the nationalist movements of the time. Rather it is to recognize the bourgeoisie as a progressive force in this period, and to recognize that this class succeeded in generalizing its own agenda as a broader docket for social change, directing historical change toward a national division of economic, political and social interests.

Likewise today, as that system of national states no longer serves the purposes for which it was established, the struggle over the political geography of a "New Europe" involves clear social divisions of class, gender and race entwined within and between the filaments of national interest. After 1992, when the European Community took a further step toward economic unification, declaring in theory if not always in practice that neither capital nor labour mobility would be restricted by national boundaries, "integration" became the watchword in public debate over the "New Europe." With unification of Europe now seemingly a reality, public discussion, mirroring the discussions in Brussels and in every national capital, turned to methods and strategies of "integration." This discussion in turn has revolved around the different ways of retaining local and "national" competitiveness within a much enlarged "postnational" territory, an estimate of winners and losers in distinctly national terms, and unabashedly nationalist jousting over particular treaties, agreements, clauses and exceptions.

The language of "integration" is self-evidently technocratic. It takes as its starting point the technocratic optimism of Maastricht, namely the assumption that European unity is a virtually accomplished fact and an unquestioned good leaving only the details to be managed into place. But more important, the integrationist perspective takes as its unexamined point of departure the perspective of capital. Whether this treaty or that treaty is preferable, whether this clause or that clause should be included – these have become questions to be arbitrated according to national economic advantage insofar as the economic elites of the different countries cannot immediately agree or, insofar as ruling class unity prevails, they are issues to be resolved as instances of national cooperation. The Maastricht agreement was a clear case of the latter in which means of economic, financial and political integration were agreed by the functionaries of the twelve member nations without serious consideration of the political opposition that such "integration" would face among the populace. As even the language of "treaties" suggests, the earlier struggle between local capitals to forge national unities is at the end of this century being substantially rerun at a higher scale. With the geographical core of the "New Europe" now substantially defined, "integration" is the rubric under which the new conditions of cooperation vis-a-vis competition are set. The largely unstated qualifier to "integration" is *economic* integration; integration is the solution to problems set by the unification of markets (including labour and financial markets) and conditions of production. Only at the margins is it a debate about social or cultural integration or the integration of environmental goals across Europe; indeed, local social, cultural and environmental differences are to be retained even highlighted within European unity, not only as nostalgic reminders of what differentiates Europe from the world outside, but as the primary commodities of the tourist industry.

It is hardly surprising therefore that opposition to a united Europe has focused on questions of jobs and social welfare, refusing implicitly or explicitly to accept the class-specific agenda denoted by economic integration. This opposition represents a very rational response to European unity, especially given the close connections between postnationalism and the retrenchment of national governments against the provision of social services. The latter constitute a clear attack along class, gender and racial lines against those populations most dependent on social services but who have been pushed to the periphery of European unity. And yet at the same time, this opposition is at times driven by a narrow nationalism, especially in the discussion of jobs – a localism that itself indulges an appeal to national and at times racial exclusivity. "Save British jobs" and "Save British industry" are a dangerous response to European unity and no response at all to the class specific agenda of "integration."

4.5 Pre-national Europe?

A very different kind of response to the fragmentation of nationalisms is emerging elsewhere in Europe. Although the precipitous disintegration of the Soviet Bloc after 1989 was in the first case a political event, substantially removed from the dilemmas facing Western Europe, these two sets of experiences are not now as disconnected as they might once have seemed. In the first place, the Soviet Union after 1917 and Eastern Europe to a more limited extent after 1945, represented significant if ultimately failed experiments at a postnational economic and political structure, albeit somewhat removed from direct capitalist forms of production and social reproduction. The dilemmas of the European Community in the latter part of the twentieth century were to a significant degree anticipated in 1917 even as the Soviet Union fought at the same time to escape the central political and economic dynamic of capital accumulation that would bring the situation to a head in the West.

In the second place, the return to regional, religious and variously "ethnic" definitions of nationhood that have especially marked the remaking of the erstwhile Soviet Union and Eastern Europe (but have also appeared in the west) have also brought about a remaking of scale intricately connected to the reassertion of various "subnationalisms." In this case, however, the remaking of scale is in large part a response not only to the "postnationalism" of communist government but also to the earlier national state formation of capitalist governments. The baseline for the Armenian struggle with Azerbaijan, the division of Czechoslovakia into a Czech and Slovak Republic, the multifaceted fissures between Serbs, Croats, Bosnians, Christians and Muslims in the ex-Yugoslavia, the Serbian ethnic cleansing in the genocidal war in Bosnia, the reassertion of Cossack identity in Russia and of monarchial claims in Romania – the baseline for all of these struggles lies less with the recent past than with the enforced nationalisms that took place in an earlier era. In retrospect, the 1919 Versailles Peace Treaty begins to appear as, among other things, a tragically unsuccessful enforced territorialization of many smaller European nationalisms, judged to occupy the periphery of an effectively complete system of European states by 1919. It is this same core, of course, that is disintegrating in the opposite direction with "European unity."

In this sense, the reassertion of pre-national regional identities in Eastern Europe (nationalisms that predate the nationstate, that is) also represent a competitive response not just to European unity but to a longstanding globalization of capital. In most cases, pre-national regionalism has taken the form of a particularly reactionary reassertion of local identities. Nor is it restricted to Eastern Europe, as for example the case of Scottish nationalism makes clear. The visceral emotional appeal to Scottish nationalism over the last two decades may not have achieved such a politically charged or as violent a reaction to globalization and enforced nationalism as is now evident in much of Eastern Europe, but it springs from many of the same concerns. It is every bit a pre-national reassertion of regionalism vis-a-vis a British state officially dominated by England since the early eighteenth century, and is rooted in a quite mythical history of national unity prior to Union. What distinguishes Scottish nationalism in recent years is the somewhat novel attempt to use the "New Europe" as an appropriate venue for advancing its claims, thus vaulting over the scale of national government. That at least might be considered progressive. Such a tactic is only partly successful, of course. The central national government in London has not only limited severely the extent to which it will allow the borders of Britain to be made more porous. It has also in part succeeded in deflecting the challenge of Scottish nationalism into a marketing device; the kilt, Scotch whisky and the Scottish Highlands are now ready symbols for a *British* tourist industry capitalizing upon "regional diversity." The same point applies to the repackaging of Breton claims of national difference into celebratory symbols of *French* uniqueness. As the nation goes multinational, the region in different ways makes claims for the national (jumping scale) or is made to stand in for it.

4.6   Politics, Geography and the Periphery

Marx concluded the first volume of *Capital* with a chapter on colonization which, though much misunderstood, makes the powerful point that the social relations of capitalism are more clearly and sharply observable at the periphery of the system than at the centre. The same lesson may well apply in the context of a "New Europe" and the "integration" which is being led from the conservative core of Brussels and Bonn, London, Paris and the Hague. So evident is this locus of power that, as the title of this volume suggests, peripheral "regions", from the Celtic and Nordic to the Eastern and Southern European find themselves forced to consider competitive responses. It would be a mistake, however, to accept the terms of this challenge in the national and often nationalist garb in which they are dressed. As the Scottish example suggests, it is at the periphery where the contradiction between a prenational and a post-national Europe is most intensely felt. On the one hand, a pre-national regionalism is not only quite unrealistic insofar as fragmented prenational regions would lack the power to compete under the economic conditions that have led in the first place to the diffusion of the power of national state boundaries and the entertainment of transnational possibilities. But on the second hand, it is difficult to imagine that such a pre-national regionalism could be a progressive historical force. Rather, such a fragmentation would in all likelihood set off a dangerous economic and cultural competitiveness between regional states defined through their appeal to a quintessentially *local* exceptionalism buttressed by historically partial reinscriptions of past calumnies.

If pre-national regionalism is no realistic option, nor is simple acquiescence to the intensified peripheral exploitation and marginalization that is likely to result from integration into a post-national Europe.

Much as Marx argued about European-held colonies, the sharp differences of interest (in class, race and gender terms) pertaining to European unity are most sharply visible at the periphery where the conservatism of core governments is less commanding. And yet the social democratic opposition to the European Union, which pays lip service to such social differences, has hardly been more realistic. Swedish and Finnish integration into the European Union is indeed likely to involve a dilution or even dissolution of progressive social welfare systems (especially affecting women), somewhat protective labour legislation (affecting workers, especially migrant workers), and emerging environmental gains in those countries; it will also extend the penetration of privatization and commodification into the rhythms of daily life, much as social democrats have argued. The unspoken corollary of this prediction, however, is that it will be difficult for peripheral capitalisms outside the EU to sustain their previous systems of social services. Privatization and the dismantling of the social welfare systems in Britain or Sweden may have been the work of especially ideological conservatives, but they also represented a response to economic globalization, post-nationalism, and the precarious position in which national state found itself. Even social democratic governments outside the EU have been forced to initiate or sustain austerity cuts. Reversing this movement will take a more profound political shift than that implied by inclusion in the EU.

In a quite pessimistic assessment of Europe and the Left, David Marquand's (1994, 24–26) only allowance to optimism is that at least European Unity is controlled by political rather than economic means and this may afford the Left some kind of entre. Too many on the Left, he says, have put "the cart of economics" before "the horse of politics":

> There is, in short, a contradiction between the monetary ambitions of the Union [for monetary unity] and its territorial divergences. Unless and until that contradiction is resolved, the Union is as likely to move backward as forward. And the contradiction can be resolved only by and through political institutions.

But it may be Marquand who has the cart of a defeated pessimism before the horse of politics. What else is one to make of his resignation to the idea that the contradiction between economics and geography will only be resolved via the institutional structures of the Union? For no other alternative is proposed. In the first place, as the earlier discussion of scale suggests, postnationalism may not be inevitable but it is backed by a powerful logic of the political economic geography of capital accumulation which, far from being comprehensible in terms of forward and backward, is multifaceted. Second, in recognition of this historical restructuring in the geographical scales of social organization, an internationalist response to the ongoing formation of a new Europe is imperative. But such a response involves a more ambitious mix of *political* competition and cooperation than Marquand envisages: competition to ruling agendas for economic restructuring, guided first and foremost by the dictates of profitability; cooperation across national boundaries between political movements with an inherently *social* (rather than economic) vision of different possible futures.

If, as many geographers have been arguing for two decades, it is appropriate to see social geographical forms as the temporary fossilization of social relationships, and vice versa, then the reterritorialization implied by a New Europe must be seen as equally a new socialization in which reinscribed relations of class, race and gender are re-etched as new structured geographies of social difference. That the accomplishment of a New Europe is likely to lead to an exacerbation of existing patterns of uneven development, notwithstanding new pockets of development taking advantage of cheap labour and EU subsidies applied to the poorer periphery is a more appropriate entre for an internationalist Left opposition than existing political institutions. Where Marquand's pessimism may be intuitively justified is that the progress of European unity on a broadly capitalist basis has dramatically outstripped any internationalist response from feminist and green, socialist and anti-racist movements.

The size of this task should not be underestimated. The power of the ruling classes to dictate the agenda for European integration has been considerable; they have already jumped scales leaving any possible internationalist opposition to begin almost from the beginning. They have successfully cast their own economic interest for profits in terms of jobs and thereby significantly fragmented increasingly conservative and isolated union movements along national lines. That the task now of asserting popular interests seems so great is in many ways the direct result of the missing networks of crossnational grassroots political organization. To the extent that such modest networks now emerging can have a significant effect on the conditions of European integration, they will express popular social interests in global interconnectedness rather than the economic interests of ruling elites. A more egalitarian social geography of Europe should be the political goal of these movements.

References

Anderson, B. (1983), *Imagined Communities*, Verso, London.
Appadurai, A. (1993), 'Patriotism and its futures', *Public Culture* 5, 411-430.
Froebel, F., J. Heinrichs, and Kreye, O. (1980), *The New International Division of Labour*, Cambridge University Press, Cambridge.
Gupta, A. (1992), 'The Song of the Nonaligned World: Transnational Identities and the Reinscription of Space in Late Capitalism', *Cultural Anthropology* 7, 63-79.
Herod, A. (1991), 'The Production of Scale in US Labour Relations', *Area* 23, 82-8.
Herod, A. (1992), *Towards a Labor Geography: The Production of Space and the Politics of Scale in the East Coast Longshore Industry, 1953-1990*, Unpublished Dissertation, Department of Geography, Rutgers University.
Hobsbawm, E. (1990), *Nations and Nationalism Since 1780*, Cambridge University Press, Cambridge.
Jonas, A. (1994), 'The Scale Politics of Spatiality', *Environment and Planning D: Society and Space* 12, 257-64.
Mandel, E. (1975), *Late Capitalism*, New Left Books, London.
Marquand, D. (1994) 'Reinventing Federalism: Europe and the Left', *New Left Review* 203, 17-26.
Marston, S. (1990), "Who are the People'?: Gender, Citizenship and the Remaking of the American Nation', *Environment and Planning D: Society and Space* 8, 449-58.
Marx, K. (1867), *Capital* vol. 1., International Publishers (1967 edn.), New York.
Murray, R. (1971), 'Internationalization of Capital and the Nation State', *New Left Review* 67.
Nairn, T. (1977), *The Break-Up of Britain*, New Left Books, London.
Paasi, A. (1991), 'Deconstructing Regions: Notes on the Scales of Spatial Life', *Environment and Planning A* 23, 239-56.
Piore, M and Sabel, C. (1984), *The Second Industrial Divide*, Basic Books, New York.
Rowthorn, R. (1971), 'Imperialism: Unity or Rivalry?' *New Left Review* 69.
Saegert, S. and Leavitt, J. (1990), *From Abandonment to Hope*, Columbia University Press, New York.
Smith, N. (1984), *Uneven Development: Nature Capital and the Production of Space*, Basil Blackwell, Oxford.
Smith, N. (1992) 'Geography, Difference and the Politics of Scale', in J. Doherty, E. Graham and M. Malek (eds.) *Postmodernism and the Social Sciences*, Macmillan, Houndsmills, 57-79.
Taylor, P. (1981) 'Geographical Scales in the World Systems Approach', *Review* 5, 3-11.

CHAPTER 5

Cross-border Co-operation and European Regional Policy

Anne van der Veen
University of Twente

Dirk-Jan Boot
University of Twente

5.1 Introduction

Cross-border co-operation between local and regional authorities is increasingly attracting attention from both governmental practice and (social) science. The reasons are twofold. First, there is a growing need for information about how to organise co-operation. Secondly, some insight into the functional foundation of co-operation is necessary: What is the reason for co-operation and what is the content? This paper will focus on both aspects of cross-border co-operation. A distinction will be made between internal regions in the European Union (EU) and regions external to the EU.

As far as both external and internal regions of the European Union are concerned, the general question is the same with regard to the organisational aspects of cross-border co-operation: What level of government in cross-border regions should supply which kind of public or merit goods, such as infrastructure and environmental protection? Standard literature on fiscal federalism for *national* organisational problems discusses the contrast between centralising and decentralising factors (Oates 1972). Moreover, another possible starting point for the analysis of this issue is a seminal paper by Olson (1971) on the logic of collective action. In this chapter we will apply both lines of arguments to cross-border co-operation between government layers of different countries.

In discussing the functional aspects of co-operation, there are several reasons to focus on cross-border co-operation. The important motives dealt with here are first the aim towards *regional development* and, secondly, *European integration*. We will therefore concentrate on internal borders. Internal border regions traditionally are lagging regions, a situation which often is due to a geographically peripheral position. To make up arrears for this, border regions have started to cooperate with their cross-border counterparts. Cross-border co-operation since 1993, however, also has a wider European meaning in facilitating European trade. Because these aims have a direct relation to the objectives of regional policy, the removal of regional disparities and the problem of cross-border co-operation will be linked here to the problem of (national and European) regional policy.

In section 5.2 we attend to the phenomenon of cross-border co-operation. The section contains a short characterisation of the position of border regions and presents some notions about their changing position. It also pays attention to the institutional consequences of efficient co-operation. Section 5.3 explains the rationale of national and European regional policy. We will discuss the dilemma of equity and efficiency and the problem of competencies in the design and implementation of regional policy. Moreover, some of the direct linkages between European policy and border-regions are examined. Section 5.4 contains the linkage between cross-border co-operation and regional policy. Finally, section 5.5 will present an elaboration of the main argument of the paper.

5.2 Border Regions

A border region is an area adjacent to an international boundary, whose population is affected in various ways by the proximity of that boundary (Anderson 1983, 1). In the past, areas along the border were primarily nationally oriented. People were forced to keep their attention turned toward the national centre (Strassoldo 1983, 123). This led border regions to suffer from economic and social marginalisation. The needs and aspirations of border regions received little attention from distant national capitals. Issues involving adjacent border regions generally had to be dealt with by the respective national governments, because international law and diplomacy was the only way of cross-border communication (Hansen 1983, 262). In this sense the existence of frontiers undeniably obstructed the development of the border regions.

5.2.1 The Changing Position of Border Regions

Cross-border relations between countries are based upon common interests. Along the border basic matters like Customs and water management have to be arranged. In the political setting of each relation these kinds of arrangements are handled more or less easily. Ethnic and cultural relations across the border clearly facilitate the organisation of these interests.

In the last decades, however, the position of border regions has changed. Altering social and economic circumstances has brought on a more internationally oriented way of life. European borders have become increasingly permeable to the mobility of persons, goods, services and information. First, political internationalisation between East and West Europe has created room for cross-border contacts (Veggeland 1993 and Storm Pedersen 1993). Secondly, European integration has contributed to this 'new world'. The emergence of the internal market has allowed more interdependency between nations and the enclosed regions. European countries and urban regions are more interlinked now and border regions will assume a key role. Nowadays, regions along the border that were peripheral from a national point of view become the linking regions in the integrating Europe. This applies, by definition, to the internal border regions of the EU.

For internal regions this new situation sets some claims on the region's infrastructure and spatial quality (Cappellin 1991 and Van der Veen 1993, 90). Important factors for economic development are:

- presence of gateways between the national infrastructure systems;
- the location of nodal points, where different types of transport meet, and where public and private services concentrate;
- alliances between regions, resulting in clear-cut government policy regarding physical planning, environmental policy and regional planning.

However, especially in these fields, border regions are confronted with their special situation. In the discussion about border regions, their deficient infrastructure is a frequently mentioned problem. Border regions are handicapped by inadequate cross-border infrastructure, due to the historical separation of national states (Hitiris 1991, 240; Rietveld 1993). More generally, spatial quality suffers from the proximity of a border. There is no all-comprising planning and co-ordination institution. Each country has its own planning procedures. Differences in physical planning and in the legal status of that planning lead to conflict between nationally pursued spatial developments and those strived for in neighbouring countries. These conflicts will undoubtedly be most problematic in border regions. Besides, nationally oriented planning will often cause negatively valued land-uses (nuclear power stations, disposal sites for radioactive waste, rubbish storage) to concentrate near a border (Sayer 1983, 68).

5.2.2 Cross-border Co-operation

The new position of border regions offers an explanation for the rise in cross-border co-operation between local and regional authorities. It is an attempt to solve the problem of regional underdevelopment by taking the opportunities created by the political internationalisation and European integration. Whereas this argument is functional by nature, there is also an organisational reason: Local and regional authorities along the border become aware that their individual efforts will not be very effective. They realise that a co-operative approach would better fit the situation. An individual attempt to exploit their changing position may discharge into a strong competition between cities and regions (Cappellin 1991) in which presumably most of them will lose. This is because of the nature of interests which are at stake. To model this behaviour we will discuss the theory of fiscal federalism that deals with the division of labour between government layers and theories regarding co-operation between local governments.

Fiscal Federalism and the Logic of Collective Action

Within the boundaries of a *nation* each country has its own internal solution to the dilemma of centralisation and decentralisation (Oates 1972; Van der Veen 1993, 87). This dilemma consists of the confrontation of decentralising the provision of public and merit goods to the level of government which best fits to the tastes of consumers (given their preferences for goods and taxes, Boadway and Wildasin 1984) and, on the other hand, centralising government tasks on a administrative high level, because of economies of scale and *external effects*. Each country finds its own administrative structure with, for example, the

Netherlands as a highly centralised country and Germany as a federal state.

This organisational equilibrium is, of course, not fixed and static. There is room for modifications. First, local governments may try to escape from centralising forces by horizontal co-operation. Olson (1971) describes the favourable circumstances for such a deliberate agreement. Economies of scale will ease the completion of such a co-operation but external effects may hinder the realisation (See also Denters 1987, 127-148). Secondly, the equilibrium may change due to external forces. Factors such as the internationalisation of the economy for instance, may potentially change administrative structures of a country.

The question now is whether notions of fiscal federalism and collective action can also be applied to the situation of *cross-border* co-operation. Following the theory there is ground to equip cross-border regions with those tasks and competencies which cannot be dealt with on a (local) interior level and which need not necessarily to be dealt with on a (supra) national level.

The problem which deserves attention here is that a specified competence structure will only be of interest as far as it facilitates or improves the function it addresses. There is a lot of discussion about the connection of *administrative scale* and function. In this debate economic aspects and political values play a role. The economic aspects concern efficiency arguments, the political values concern democracy and legitimacy. Although there seems to be no identical idea of an optimal administrative scale, both economists and political scientists are aware of the fact that a major dismantling of administration is not the best solution. From an economic point of view a patchwork pattern of administration structures will cause enormous decision-making costs and face the problems of externalities and missing economies of scale (Oates 1972). From the political point of view this would cripple effective decision-making on supra-local matters and cause problematic policy co-ordination.

Besides the scale of administration, *competence* is also an important factor. As far as cross-border authorities exist, autonomous or as a consultation-body of composing authorities, these are confronted with the problem of lacking competencies to affect measures which can further unification of the region and its welfare. They have little impact on regional development. Even the construction of small-scale infrastructure is not possible without intervention and exertion of higher governments. The smaller the regions are, the more acute the problem. Until recently, neither national nor European legislation dealt explicitly with the possibility of cross-border administrative structures.

Applying Olson (1971), this dilemma may, under particular conditions, force authorities to cooperate. In recent literature the logic of collective action has been applied to situations of international co-operation (See e.g. Martin 1994, 478). National states face collaboration dilemmas in case of self-interested behaviour, or co-ordination dilemmas in case of distributional conflicts. The design of institutions is in both cases strongly influenced by heterogeneous interests and capabilities. As far as we know there is no literature on how local governments design institutions regarding cross-border matters, given the problems national governments have in collaborating or co-ordinating. It is doubtful if an individual local (or regional) authority can bring about important changes in, for instance, spatial structure. Competencies on this matter are usually situated on a higher level of government, precisely because of the far-reaching spatial consequences. This justifies the conclusion that *spatial external effects* are an important aspect in the discussion about the rationale of cross-border co-operation.[1] This argument is perhaps one of the most commonly advanced. As Anderson (1983, 3) poses:

'As social and economic activities spill over the frontiers or their consequences come to be strongly felt across the frontier, different levels of transfrontier political and administrative co-operation become necessary'.

Present-day Situation

Looking at the present-day situation of cross-border co-operation in Europe, we perceive a patchwork pattern as Oates (1972) already observed, consisting of overlapping transborder structures. Along the Dutch borders, for instance, tens of initiatives for co-operation between local and regional governments already exist, and undoubtedly new ones are being developed. Mapping out all initiatives, a pattern of geographically overlapping jurisdictions arises, varying in scale and scope. In

---

[1] For an old-dated but still timely treatise of the problem of bordercrossing externalities see Sayer (1983).

terms of administrative efficiency and democratic control this situation seems to be far from optimal. There is however no limiting power yet, which can stop the emergence of new transborder structures.

Centralising factors are frustrated because:

- It is not possible to replace existing inner-state administrative entities with a new cross-border entity. The existing national legal systems do not meet this situation, and therefore each new cross-border body is supplemental to the existing ones. We do not see that the answer of functional federalism, by Casella and Frey (1992) in their discussion of overlapping political jurisdictions, will give a solution.
- Competencies of existing local and regional authorities on both sides of the border often diverge. For most problems intended to be dealt with on a cross-border level, another partner has to be found.

These factors lead to an increasing amount of cross-border structures. In this sense, cross-border co-operation implies a threat to itself. Cutting up of governmental tasks would have enormous decision making costs and suffer from democratic fallacies. In order to prevent the downfall of what has been reached, a *first conclusion* is that more intervention by both national governments and the EU is absolutely necessary. The aim of this should be to establish an efficient administrative structure in border regions.

External and Internal Border Regions

Having discussed the rationale for cross-border co-operation we will return to the difference between external and internal border regions. There clearly is no difference in the rationale for cross-border co-operation between external and internal regions. The theoretical structure is the same. This implies that for both kinds of regions more intervention by national governments is necessary. The context in which the co-operation takes place, however, is different. The EU is an additional layer of government which may influence cross-border co-operation. This can be seen, for instance, in the case of the Nordic Countries which face a rather complicated situation. Norway and Iceland have decided not to enter the EU, whereas other countries are in a situation of transition. The Baltic states are at the moment external to the EU, but are expected to join the EU at some future point. Moreover, Norway, Sweden and Finland already have a rather intense form of cross-border co-operation, for instance with regard to labour market issues. Cross-border co-operation between Finland and Russia, however, starts almost from scratch, whereas there are relatively strong ties between Finland, Sweden and the Baltic states. In the next section we will discuss the EU as an additional complication for cross-border co-operation.

5.3 Regional Policy

As noted above one of the reasons for cross-border co-operation is to make up arrears. There thus seems to be a direct relation between the efficiency and equity argumentation which determines the logic for designing regional policy (Okun 1975). In this section we will concentrate on the problem of designing regional policy for *internal regions* of the EU. We therefore discuss national and European government activities which attempt to solve the problem of uneven regional development. Special attention will be given to the question of co-ordination and competencies.

Moreover, cross-border co-operation may have considerable external effects on other regions. This influences the targets European and national governments set with respect to regional policy and thus influences the efficiency and equity assumptions of regional policy. This second issue will be dealt with in the section 5.4.

5.3.1 National Regional Policy

It is not within the scope of this chapter to extensively dwell on the *assumptions* of regional policy in the different European countries. It suffices to remark on the trunk-lines. Regional policy is traditionally linked to the trade-off between the concepts of efficiency and equity (Okun 1975; Molle 1991; Armstrong and Taylor 1993). These two elements usually are seen as antipodes. Emphasising one will be detrimental to the other. Efficiency arguments in designing regional policy favour the strong regions in a country, assuming that the less developed regions will benefit in the long term by experiencing spillovers. The concept of equity stresses the element of income distribution in public policy. Differences in income and unemployment are the main parameters to design instruments. Every country in the EU has its own mix of efficiency and equity.

5.3.2 European Regional Policy

European regional policy is a rather new phenomenon. Until 1975 there was in fact no distinct regional policy on a European level. Concern about the possible impacts of a barrier-free Europe led to the design of European regional policy (Anderson 1993).

There is an extensive literature about the ultimate objective of the project 'Europe'. Hitiris (1991, 44-51) argues that articles of the Treaty of Rome, subsequent agreements and declarations of Heads of State, and the course that the European Community (EC) has hitherto followed, show clearly the aim of the contracting partners: realisation of 'an ever closer union among the European peoples'. The process containing this aim is usually referred to as European unification. Therefore economic integration is a prime parallel route towards the 'desirable objective' of establishing a European union. Economic integration is however not the only aspect of the EU. There are also some more politically and socially slanted aspects, which have become more important as the single market project approaches completion. Some of these aspects take shape in the (social) cohesion-thought.

European Integration and Efficiency

The main objective of European economic integration is to enhance allocational efficiency of the economies of the member states by removing barriers to movement of goods, services and production factors (Molle 1991, 361). Abolition of trade restrictions between members brings about immediate general benefits to the participants, the so-called static economic effects. These effects are derived from increased competition and trade with the existing structure of production. In the longer term, rationalisation and competition within the enlarged market are expected to accelerate development and to increase welfare because of the dynamic effects of economic integration. They consist of:
- improvement in allocation and utilisation of resources within and between participating countries;
- specialisation according to comparative advantage;

- realisation of economies of scale in both production and demand (Hitiris 1991, 5).

It may be obvious that these effects cannot fully be taken advantage of via the efforts of one country on its own.

Inequality as a Consequence

In pursuing efficiency there is a chance of causing inequalities. As Molle (1991, 417) postulates: 'There is a general assumption, if not conviction, among economists and politicians that competitive markets (efficiency) generate considerable inequality'. These inequalities are caused by structural changes which economic integration generates, like the relocation of economic activities or the changing composition of sectoral activity.[2] So, structural changes have consequences which are more disadvantageous for some groups than for others. These groups can be defined socially (particular groups of labour force) or geographically (regions or countries).[3] Here we will especially pay attention to regional disparities and not to the disadvantages to social groups.

With respect to the geographical viewpoint, Hitiris (1991, 232 vv.) argues that regional growth tends to be concentrated in 'poles of development', that is in geographic areas providing new investment with economies of scale, thus making it possible for them to gain an initial headstart and to continue to grow at the expense of other regions of the economy. Furthermore, he mentions two reasons why the process of economic integration is likely to intensify already existing problems of regional disparities.[4]

- Economic integration, free trade, enhanced competition and freer mobility of factors of production will tend to equalise

---

[2]Armstrong and Taylor (1993, pp. 283-285) suggest that this divergence *could be* only temporary: 'It is possible that we are witnessing the temporary victory of the forces of divergence over forces of convergence but that eventually we shall return to the slow long-term narrowing of regional disparities.'

[3]Of course this is not a strict distinction. The distinguished groups can coincide. For instance in the case of border regions, it is both the region itself and the group of peoples who live there, which are afflicted by the proximity of a border.

[4]Evidence for the growing problems in peripheral regions of the EC can be found in: Commission of the European Communities (1990).

commodity and factor prices. However, productivity differentials will continue to exist and they will favour the technologically advanced firms of developed areas within the economic union.
- Economic integration may encourage the concentration of new industry and the relocation of existing industry in certain areas of the economic union which give superior infrastructure, lower transport costs and availability of skilled labour. Because these factors are commonly less available in regions with economic arrears, the latter will be affected negatively and disproportionately by the process of integration.

As a result the rates of growth in developed areas will be higher than in less developed regions of the economic union, implying an unequal distribution of the gains of integration. Obviously this is not favourable to the furtherance of economic and political cohesion within the economic union. Groups that understand they are apt to be losers in the integration game, may be inclined to step out (Molle 1991, 164). This situation, of course, is most threatening when a country as a whole likely is to suffer from integration. Deprivation of a smaller area can complicate the progressing integration as well, by stimulating nationalistic tendencies. To avoid any delay of the integration process, the European Commission has taken upon itself to reduce (relative) inequalities resulting from economic integration. Redistribution-programmes are intended to compensate suffering regions for their welfare losses. This idea has recently been strengthened and developed in the Maastricht Treaty.[5] The furtherance of social and economic cohesion seems to be a major issue for the 1990s (Begg and Mayes 1993).

Thus, regional inequality at a European scale has led to common regional policy. A purely national approach will be incompatible with the integration agreement and secondly it will be ineffective under the increased interdependence of an integrated Europe.

European Regional Policy as a Remedy

From early on, the EC has been aware of the problems caused by the process of integration. The Preamble of the Treaty of Rome already

---

[5]Treaty on European Union, Article 2.

mentioned the problem of regional disparities: member states agreed about their endeavour to:

> '...strengthen the unity of their economies and to ensure their harmonious development by reducing the differences existing between the various regions and by mitigating the backwardness of the less favoured.'

Until the mid 1970s there was however no serious regional policy, partly because the Guarantee Section of the European Agricultural Fund (EAGGF) took up almost whole the Community budget.[6] Therefore regional policy was mainly taken on by national governments. Because state aid could be in conflict with the objectives of unification, and in this way meant a direct violation to the foundations of the EU, at the time, these national policies required Community approval.

Although there was no active regional policy, it was expected that other, not exclusively regional, common initiatives would have certain positive effects on the regional problems. In the first place institutions and funds like the European Coal and Steel Community (ECSC), the European Investment Bank (EIB) and the European Social Fund (ESF) were expected to have some regional impacts. Secondly the guidance section of the EAGGF, serving restructuring and modernisation of agriculture, would also have some regional impact.

Economic recession and enlargement of the EC in the 1970s accomplished a more active regional policy at European level. In 1975 the European Regional Development Fund (ERDF) was set up. The fund was mainly used to support national governments in their efforts to reduce regional disparities. Therefore the fund was shared among member states on the ground of national quotas. These quotas were based on the seriousness of the state's regional problems.

---

[6]In the scope of this article it is interesting to note that this Guarantee Section causes some inequality as well. It artificially upholds food prices on the internal market, implying extra costs for regions with little agriculture of their own. Besides, already prosperous agricultural regions will have the greatest benefit because of the types of crops cultivated, and the scale of producing (Armstrong and Taylor 1993, 283; Molle 1991, 423).

In 1985 and 1988 the ERDF was reformed.[7] Member states no longer receive their national quotas automatically. All the resources of the ERDF are allocated on the basis of ranges, defined by a set minimum and maximum for each member state. Requests for funding have to fit in with common specific criteria. In addition, the new system is based on additional financing. Subsidies are meant to stimulate national investment in economic activities, and not to substitute for them. Therefore financial support will only be given if member states contribute to the total costs of a project as well.

In addition, the European Commission started some programmes and pilot-projects, based on Community initiative rather than member state initiative. The criteria that were used to establish the need for regional support, include per capita income and rate of unemployment.

Contemporary Regional Funds

Since 1985 the Single European Act contains an explicit obligation to pursue a policy of redistribution. Article 130A of this Act reads:
'In order to promote its overall harmonious development, the Community shall develop and pursue its actions leading to the strengthening of its economic and social cohesion. In particular, the Community shall aim at reducing disparities between the levels of development of the various regions and the backwardness of the least favoured regions...'.
With the reform of the Structural Funds introduced in 1988, this general expression was worked out in five priority objectives for the three European Structural Funds (ESF, EAGGF-Guidance, ERDF). Two objectives are primarily social and are financially supported only by the ESF. They are meant to fight long-term unemployment and to encourage young people toward working life. Three objectives are explicitly geographical and are financially supported by the ERDF, the EAGGF or a combination of the three funds.[8] The first is concerned with development of those regions which are judged to be lagging behind in their

---

[7]Reforming the fund had become necesssary because of the enlargement of the EC with relative weak countries such as Greece, Spain and Portugal. The 1985-reform introduced provisions for integrated operations and coordination of national policies.

[8]For a more extensive description of these priorities see: Begg and Mayes (1993).

economic development, designated mainly on the basis of GDP per head (Objective 1). The second aims at restructuring industrial regions in decline, which are designated on the basis of the level and rate of change of unemployment and industrial structure (Objective 2). The third is meant to stimulate restructuring rural areas which have an excessively high level of income to be part of Objective 1, but which face difficulties in development (Objective 5b).

Financial aid on the base of aforementioned objectives can only be obtained with the intervention of national governments. There is, however, a possibility to obtain more direct regional support from the European Commission. A (small) share of the ERDF is not committed to the five objectives. Article 10 ERDF-regulation[9] allows the Commission to support regions which cannot be classified within the framework of objectives. Both objective-related aid and support on the basis of Article 10 ERDF, is only possible supplemental to regional expenditures of the member states. There is a system of co-financing up to a certain percentage of total expenditures.

5.3.3 Co-ordination of National and European Regional Policy

At this moment national and European regional policy exist side by side. National regional policy by member states is shaped by the choice between equity and efficiency, while European regional policy is framed by equity considerations to mitigate the impacts of integration. This implies the question on the policy level which is expected to take action to fight against regional disparities. A strong case can be made for establishing and strengthening European regional policy (Armstrong & Taylor 1993). This does not mean that regional policy should be controlled entirely from the centre and that member states should give up their own regional policies. Such an approach implies a threat of a uniform solution to solve problems which by their very nature differ from one region to another. The substantial advantages of local knowledge and experience in designing particular regional policy programmes should not be ignored.

---

[9]Commission of the European Communities (1988) Council Regulation No. 4254/88 of 19 December 1988.

As a corollary, analogous to the treatment of fiscal federalism with regard to national and local government tasks, a quick conclusion could be that the same analytical framework be applied to the division of labour between Europe and the member states. The balance between centralising and decentralising forces, incorporating external effects and economies of scale, would build the ultimate government level for performing regional policy. In modern Euro-language, this balance is translated with the term *subsidiarity*, which means that the EU should do only what member states cannot do themselves. Member states and not the EU should act, unless a good argument can be made against this practice. This principle, however, does not fully solve the problem of demarcation. Moreover, what makes the conclusion even unlikely is the fact that the EU has direct contacts with local governments which bypass national governments. This is especially the case for cross-border co-operation. We will move to this subject in section 5.4.

5.4 Cross-border Co-operation and Regional Policy

The objectives of cross-border co-operation and regional policy are closely approaching each other. As we have seen, cross-border co-operation is aimed to get rid of lagging positions by utilising new opportunities. European regional policy is meant to eliminate existing and arising inequalities between different regions. Because border regions have specific disadvantages compared to other regions, the reasons for the EU's partial direction of regional policy at border regions is obvious.

Structural Funds are distributed among regions in extensive deliberation with the national governments of member states. The European Commission, however, pursues *a more direct relation with regions* and big cities in Europe. This appears amongst others from the establishment of the Committee of the Regions, as a result of the Maastricht Treaty, and from the Delors-II Report.[10] Looking at the great number of lobby-offices of subnational authorities in Brussels, the strive for more direct contact seems mutual. Border regions are seen as areas of special interest, because of their experimental function for Europe. Border regions are the experimental gardens for continuing integration.

---

[10]Commission of the European Communities (1992), COM(92) 2000.

Therefore the European Commission attaches a high value to co-operation between local and regional authorities of different member states. The Commission expects cross-border co-operation to develop better insight into European matters, and to cultivate a growing interest for European policy. This is emphasised by the placement of a large sum of disposable funds beyond the regular objective bounded Structural Funds.

5.4.1 EU Financial Support to Cross-border Regions (INTERREG)

Many border regions come under one or more of the objectives of the Structural Funds. A share of these funds is specially reserved for cross-border co-operation. However, some border regions do not correspond to the objectives. Since the Commission is convinced of the relevance of cross-border co-operation in general, the Commission makes use of their possibility to take initiatives according to Article 10 ERDF. As was already mentioned, this Article allows the support of regions beyond the scope of the regular objectives. Border region oriented funds - both the regular and the ERDF Article 10 - are integrated in the INTERREG Community Initiative, launched in 1990.[11] Like Structural Funds, this support is based on the principle of additionality. Up to 50 per cent of total project costs can be financed from this source.

The main objectives of INTERREG are:
- to help areas within the EU along the internal borders make up their structural arrears, springing from the relative isolation of national and European economic centres, and
- to stimulate completion and development of networks of cross-border co-operation.

These objectives again show the coincidence of regional policy and cross-border co-operation.

5.4.2 External Effects, Equity and Efficiency

Cross-border co-operation is intended to eliminate disadvantages which border regions perceive with respect to regions not adjacent to an international border. In this conception cross-border co-operation is based

---

[11]Commission of the European Communities (1990), Official Journal C215 of 30.8.90, no. 90/C 215/04.

upon the equity argument. This explains the linkage to (European) regional policy. As was stated above, in the design of regular (national) regional policy there is always the trade-off between efficiency and equity considerations. Emphasising one of them is detrimental to the other.

However, the external effects of cross-border co-operation change this contrast. In cross-border co-operation there is a chance for these two elements to coincide. The changing position of border regions requires attention to the physical and spatial structure in those regions, for instance for environmental policy and cross-border infrastructure. Realising that the existing infrastructure is the (partial) cause of present-day arrears, an improvement in cross-border linkages creates an opportunity for regional economic growth in peripheral areas (*equity*). As an external effect, improved cross-border linkages will also result in improved linkages between the growth poles of the European network of urban regions. 'Better linkages' also means shorter distances or faster movements. In economic location theory both imply an enhancement of *efficiency*, which at first instance is favourable to the core regions of a country. Creation of new linkages could also imply a discharge of existing linkages. As the latter get overburdened, new linkages would have positive effects on both capacity and environmental inconvenience. This can be seen as an *efficiency* argument as well.

From a theoretical point of view one of *the interesting corollaries of cross-border co-operation* is the coincidence of the equity and efficiency arguments in regional policy, because of the potential spatial external effects of cross-border co-operation.

5.4.3 Cross-border Co-operation, Regional Policy and Competencies

Although equity and efficiency coincide there is the remaining problem of competence concerning regional policy which was discussed in section 5.3. In fact it is the problem of national sovereignty. In the early literature on cross-border co-operation, Hansen (1983) argued that completion of cross-border co-operation depends upon the vehemence of sovereignty values. If these values are dominant, transboundary co-operation will not arise, whatever the economic as well as social and cultural advantages may be. This may be seen as the conflict between economic and political values. Nevertheless, current practice demonstrates a clear call for cross-border co-operation, and the economic rationale is obvious. Sovereignty values are presumably not dominant in the EU at this moment. It has to be recognised however that there are some signals that an opposing attitude is dominating in some of the EU member states.

At the moment, for instance, two treaties between The Netherlands and its adjacent countries exist, which make it possible to start transborder structures based on public law. These treaties can be seen as an important replenishment of the limited usefulness of private law. The purport of the treaties shall not be discussed here. It suffices to say that there is some evidential progress made on this subject.[12] However, the comments of the national parliaments and concerned public institutions regarding ratification of the bilateral treaties between The Netherlands and Germany and between The Netherlands and Belgium, clearly demonstrate that the integrity and sovereignty of member states is at stake. The direct contacts between the EU and cross-border regions are an example of a lack of demarcation of competencies.

5.5 Evaluation and Outlook

Due to European integration and political internationalisation border regions now face new situations. By stimulating cross-border co-operation, the former peripheral regions will make up arrears. Border regions may be seen as key territories in Europe. They may link the growth poles in the European network of cities. In this chapter we discussed the economic rationale of cross-border co-operation for internal and external borders in terms of Oates' fiscal federalism and Olson's logic of collective action.

The main conclusion was that national states have to act to organise the current patchwork pattern of cross-border co-operation along internal and external borders. National and local governments must find a solution to the classic problem of optimal supply of public goods. Sec-

---

[12]There are however some serious restrictions to the practical utility of the treaties. This appeared from a symposium in Papendal, The Netherlands 9-10 December 1993, on the occasion of the 'birth' of the first cross-border region (Euregion Rijn-Waal) based on the treaty between The Netherlands and Germany.

ondly, in accordance with the first conclusion, competencies in a European context, with respect to national regional policy and European regional policy will have to be demarcated. Thirdly, as the main corollary, efficiency and equity arguments in regional policy coincide as cross-border infrastructure and cross-border physical planning in the European Union are beneficial for cross-border peripheral regions, as well as for core regions in the European member states.

References

Anderson, J.J. (1993), 'Skeptical reflections on a Europe of Regions: Britain, Germany and the ERDF, *Journal of Public Policy*, 12, 417–447.
Anderson, M. (1983), 'The Political Problems of Frontier Regions', in Anderson, M. (ed.), *Frontier Regions in Western Europe*, Frank Cass, London.
Armstrong, H. and Taylor, J. (1993), *Regional Economics & Policy*, Harvester Wheatsheaf, Hertfordshire.
Association of European Border Regions (1991), *Linkage Assistance and Cooperation for the European Border Regions (LACE) Cross-border Cooperation in Practice: Infrastructure and Planning in Cross-border Regions*, Euregio, Gronau.
Begg, I. and Mayes, D. (1993), 'Cohesion in the European Community: a Key Imperative for the 1990s?', *Regional Science and Urban Economics* 23, 3, 427–448.
Boadway, R.W. and Wildasin, D.E. (1984), *Public Sector Economics*, Little, Brown, Boston.
Cappellin, R. (1991), 'Interregional Cooperation and Internalisation of Regional Economies in Alpe Adria', Paper presented at the 31th RSA European Conference, Lisbon, Portugal.
Casella, A. and B. Frey (1992), 'Federalism and clubs; Towards an economic theory of overlapping political jurisdictions', *European Economic Review*, 36, 639–646.
Commission of the European Communities (1991), *The Regions in the 1990s, Fourth Periodic Report on the Social and Economic Situation and Development of the Regions of the Community*, Office for Official Publications of the European Communities, Luxembourg.
Denters, B. (1987), 'Gemeentelijke samenwerking: de lappendeken in de lappenmand? Een politicologische benadering', in T.P.W.M. van der Krogt e.a. (red.), *Big is beautiful? Schaalveranderingen in overheiden samenleving*, VUGA, Den Haag.
Hansen, N. (1983), 'International Cooperation in Border Regions: An Overview and Research Agenda', *International Regional Science Review* 8, 3, 255–270.
Hitiris, T. (1991), *European Community Economics*, Harvester Wheatsheaf, Hertfordshire.
Martin, L.L. (1994), 'Heterogeneity, linkage and commons problems', *Journal of Theoretical politics*, 6, nr. 4, 473–492.
Molle, W. (1991), *The Economics of European Integration: Theory, Practice, Policy*, Dartmouth, Aldershot.

Oates, W.E. (1972), *Fiscal Federalism*, Harcourt Brace Jovanovich, New York.

Okun, A. (1975), *Equality and Efficiency: the Big Trade Off*, Brookings, Washington.

Olson, M. (1971), *The Logic of Collective Action: Public Goods and the Theory of Groups*, Harvard University Press, Cambridge.

Rietveld, P. (1993), 'Transport and Communication Barriers in Europe', in Cappellin, R. and Batey, P.W.J. (eds.), *Regional Networks, Border Regions and European Integration*, Pion Limited, London.

Sayer, S. (1983), 'The Economic Analysis of Frontier Regions', in Anderson, M. (ed.), *Frontier Regions in Western Europe*, Frank Cass, London.

Storm Pedersen, J. (1993), 'The Baltic Region and the new Europe', in Cappellin, R. and Batey, P.W.J. (eds.), *Regional Networks, Border Regions and European Integration*, Pion Limited, London.

Strassoldo, R. (1983), 'Frontier Regions: Future Collaboration or Conflict?', in Anderson, M. (ed.), *Frontier Regions in Western Europe*, Frank Cass, London.

Veen, A., van der (1993), 'Theory and Practise of Cross-border Cooperation of Local Governments: the Case of the EUREGIO Between Germany and the Netherlands', in Cappellin, R. and Batey, P.W.J. (eds.), *Regional Networks, Border Regions and European Integration*, Pion Limited, London.

Veggeland, N. (1993), 'The border region challenge facing Norden: Applying new regional concepts', in Cappellin, R. and Batey, P.W.J. (eds.), *Regional Networks, Border Regions and European Integration*, Pion Limited, London.

Part II
NORTHERN LIGHTS

CHAPTER 6

Europe of Regions – A Nordic View

Perttu Vartiainen
University of Joensuu

Merja Kokkonen
University of Joensuu

6.1 Introduction

During the middle 1980s the concept 'Europe of Regions' was advanced through public discussion and has been widely used since to give an image of the ongoing development of the united and more democratic Europe. European integration is believed to strengthen the power of regions not only in politics, but also in economy. As Caciagli (1990) has put it, "the modern Europe formed by nation states is developing into a post-modern Europe of Regions". Even though this concept contains much wishful thinking and, to be sure, political persuasion, it can be seen as a challenging starting point for serious discussion also in regional studies.

Several dimensions can be identified in the discussion of Europe of Regions. Three basic ideas to be discussed here are: 1) a federal Europe formed by *politically and administratively more autonomous regions*, 2) a Europe of competitive and co-operative *functional regions on different spatial scales* and 3) an economically strong Europe characterised by *'postfordist' industrial districts*. First, these ideas will be analysed generally and secondly, they will be handled with specific reference to the Nordic case.

From this base we shall discuss different, and maybe conflicting scenarios of a Europe of Regions and their relevance to Nordic peripheries. The evaluation is based both on our recent study on a European scale (Kokkonen & Vartiainen 1993) and the ongoing discussion in Nordic professional forums.

The term Nordic peripheries is used here to mean peripheries in a national context. In the sequel it will be characterised in terms of the relative distance from and dependency upon the national centre or centres – dependence in terms of either economy, administration or culture (cf. Rokkan & Urwin 1983). We shall concentrate on the peripheral regions of Norway, Sweden and Finland. The other Nordic states, Denmark and Iceland, are rather different, both because of their geopolitical situation and their internal geographical structure.

Regional disparities in the three Nordic countries of our focus are similar and clearly exhibited along the North-South-axis. Their peripheries have also been widely assisted by public policy incentives. Peripheral areas are characterised by a dependence upon resource-based industries, a high proportion of public sector employment, low population density, intense periods of population decline, an ageing population and high unemployment (Figures 6.1 and 6.2).[1] In addition to the natural resource endowments (timber, minerals, tourist attractions, fish etc.) peripheries have offered new human resources to core regions during the growth periods of national economies.

Nordic discussions of deepening European integration and the future of peripheries, evoke both expectations of upcoming 'Euroregions' and fears of being out-distanced by integration. These discussions have become more acute due to the recent fiscal crisis of the Nordic public economies, a phenomenon which displays the dependence of the Nordic peripheries on national welfare policies.

---

[1] After Veggeland (1993, 95) the term peripherality in a *European* context probably applies to some 10–15 per cent of the Nordic population (including Iceland in this case). The areas shown in Figure 6.1 contain 13.7 per cent of the population in Norway, Sweden and Finland.

Europe of Regions – A Nordic View 99

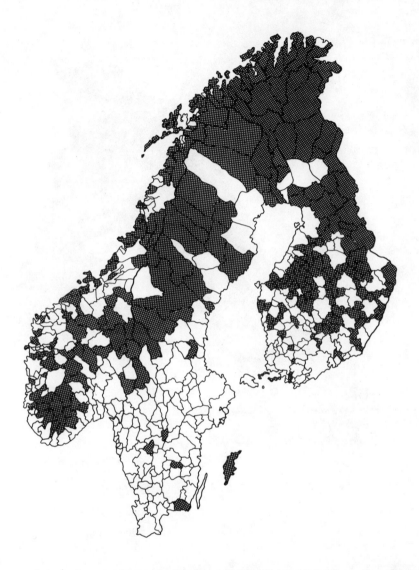

Figure 6.1. An areal representation of Nordic peripheries: the share of primary production and public service employment exceeding 45 per cent in 1990 by travel-to-work areas. In Norway public service employment is an estimate based on service sector employment.
Source: basic statistical data from national statistical centres; on public sector classifications see Regional utveckling i Norden 1993, 59.

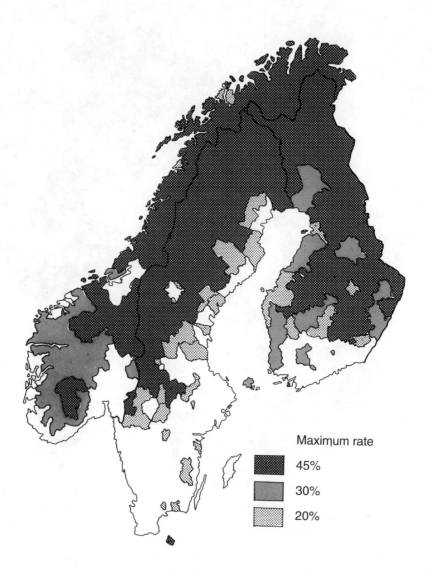

Figure 6.2. The representation of peripheries in regional policy: the assisted areas in 1994 by the maximal development aid for investments (percent), in Norway, Sweden and Finland.
Source: NOGRAN 1994, 2.

6.2 A Mosaic of European Regions

6.2.1 Regionalization as an Institutional Process

The term 'Europe of Regions' is of West European origin. Nowadays it is basically founded on the need to fill the democracy deficit of the EU. Although political and economic decisions are increasingly taking place at the international level, it has been suggested that more power be given to the local and regional level. This has been strongly advocated by the European Council and many pan-European organisations, such as the Assembly of European Regions (AER).

During the 1990s, the decision-making bodies of the EU have been more responsive to the regions' demands. The subsidiarity principle and the foundation of the Committee of the Regions have boosted hopes that the regions' voices will be heard even in the formal decision-making of the Union[2]. It must be noted, however, that many contradictory forces lie behind this tendency toward regionalization. For example, the contents of the subsidiarity principle is politically disputable – it originally referred to horizontal decisionmaking between public and private sectors, not to hierarchical central-local relations.

Strong, self-supporting regions are assumed in the basic EU ideology regarding spatial planning, especially in structural policy. The Commission has constantly stressed that the benefits of the single internal market will not be achieved without spatial cohesion. Cohesion will be achieved only by making the less favoured regions more competitive and self-reliant, thereby reducing the efficiency and productivity gap with the more prosperous regions (Amin 1992, 131). The sub-national administration has also been assumed to have greater responsibilities which has resulted in EU legislation on issues such as environmental protection and public procurement (Martin 1993, 153). These examples suggest new challenges to local and regional governments. The problems have lead to discussions about the need to reform the sub-national administration.

---

[2]The third element in giving more opportunities to regions, especially in Federal States, has been Article 116 from the Maastricht treaty, which specifies that Member States may be represented in the Council of Ministers by a regional minister in the absence of a national minister.

Several studies have recently indicated the rise of regionalization as a major trend in the administrative development of the Western European countries (see Sharpe 1992; Goldsmith 1993; Martin 1993). The concept of regionalization is usually understood as the growing regional autonomy in administration and politics. Regions are defined, most typically, as the administrative units below the nation state level and above the local or municipal level[3].

However, institutional regionalization is not a new phenomenon originating from European integration. The local and regional administrative systems vary greatly in Europe as do their territories – by size, population and the extent of autonomy in relation to the nation-state. There are nations that have chosen a strong sub-national administration as a solution to the periods of undemocratic governments. Of the member states, Germany and Belgium are usually considered to be regionalized nations and, to a lesser extent, Spain and Italy (e.g., European Parliament 1988; Charpentier & Engel 1992). These are the regions which have real power in public decision-making and finance.

These divergent sub-national structures also reflect themselves in the relations between regions and the European Union. The Europoliticians that have most eagerly advocated more powerful roles for local and regional administration have usually come from countries that already have a strong sub-national administration and an interest to preserve the status of regions in a unified Europe.

At this time, there are also signs of a growing challenge to institutional regionalization from those EU-states which have no strong regional element in their administrative system. In Great Britain, conflicting relations between the local and central government as well as criticisms about the lack of regional government have characterised broader discussions about the EU since the late 1970s (Goldsmith 1992; Hebbert 1989 and 1993; Stoker 1991). Local authorities are, on the other hand, now eager to respond to the possibilities and challenges that 'europeanisation' brings (see Goldsmith 1993 and Martin 1993). But, as Hebbert (1993) emphasises, this tendency to advocate for local solutions should not been confused with regionalism. He goes still further in unravelling the myth of British regionalism: The most obvious and

---

[3]Sharpe (1992) uses in this connection the term 'meso' (level) to avoid the confusion of using the pluralistic concept of region.

clearly defined European 'regions' in UK, Scotland and Wales, are, in fact nations.

The relation between regionalism and peripherality is often confusing. There was a new rise of ethnic regionalism and nationalism in the 1960s and 1970s in the least prosperous national peripheries. That has been explained by growing regional disparities in the economy and the political pressure of a central state (e.g., Rokkan & Urwin 1983; Keating 1988). The best examples of the strength of ideology in regionalism are, in fact, economically well-off regions, such as Catalonia and Flanders.

Nevertheless, the most powerful demand for regionalization has come from the fault-line where regionalism meets nationalism.[4] This may support the creation of new federal states, such as Belgium, or smaller nation-states, as in the East European cases. As Dréze (1993) has pointed out, deepening European integration could give those regions willing to have more autonomy a new opportunity to secede from the nation state. In principle, it could be possible to belong directly to the European Union, without having full statehood. Still the vision of nation states dissolving into a Europe of Regions or a Federal Europe remains unrealistic (cf. Keating 1993, 310). Rather, the fear of radical separatist movements may be a restraint to a deeper meaning of the Europe of Regions.

6.2.2 Changing Functional Regional Systems: Competition and Co-operation

Many scholars believe that the Europe of Regions will be more a Europe of cities or large *functional* regions because "regions without a metropolitan or city status are not able to compete effectively for employment growth, finances and greater political or administrative autonomy" (Lever 1993, 937). As cities compete they form alliances which may be both regional and transnational by nature (Cooke 1992; Nijkamp 1993). A Europe of networking regions is a scenario which *may* also give more future opportunities for other than metropolitan regions.

---

[4] Regionalism can be defined as the strive for greater territorial autonomy, not denying the legitimacy of the state. On the other hand, nationalism (or regional ethnic nationalism) aims at total separation from the nation state (Keating 1988). It must be noted that the words nationalism and regionalism may, indeed, have different meanings in different countries.

More likely it will strengthen the growth corridors connecting the main European cities.

The ideal of co-operative city regions below the metropolitan level is illustrated by the 'grape model' of Kunzmann and Wegener (1991, 64).[5] Territorial co-operation has increased and widened rapidly all around Europe, not only in the city regions. For example, the EU-REGIO-co-operation that started at the Dutch-German border in the 1960s has already spread to the Eastern borders of the EU. This form of co-operation varies from small scale intermunicipal partnership via transnational region-formation to networking between regions or cities in different parts of Europe.

Transnational co-operation, in particular, may also strengthen the position of the traditionally peripheral border regions. This being the case, the peripheral regions could take over some activities and functions which are now located in the national centres. Success in the economical development of a co-operative region is closely linked with two features, the structure of regional economies and interregional networks which exists, and the way in which co-operation can be utilised. Success needs not only friendly agreements between regional authorities, but also networking between enterprises, R & D organisations and the local civil society. Unfortunately, most contemporary initiatives toward regional co-operation have been very limited in this sense.

6.2.3 Regions as Regional Economies and Actors in a Wider Economy

The notion of co-operation always implies the notion of competition: the ultimate motive for regional agencies to cooperate is to strengthen themselves against each other. The Europe of Regions will, consequently, be a Europe of increasing economic competition in which will be found both winners and losers (Selstad 1992).

Nevertheless, the impact of local and regional administrations on economic development has been strengthening throughout the developed world, not only in Europe. The rise of local action seems connected to

---

[5]It must be noted that this model originates from Germany, reflecting the interests for a decentralized urban system as well as a federal-state structure.

economic recession, deindustrialization, widespread structural unemployment and the fiscal difficulties of the public sector since the 1970s. Consequently, ideas of market rationality and privatization have emerged. As Harvey (1989, 3–16) formulates it, the managerial approach typical to the 1960s has given way to entrepreneurial forms of local governance. Since the late 1980s 'europeanisation' has offered a new, and a more institutionalised, driving force in emphasising the actor role of local and regional governance (Goldsmith 1992; Martin 1993).

Meanwhile we have witnessed an expansion of global corporate networks (Amin & Thrift 1992) and an intensifying interplay between 'the global' and 'the local' in the new urban and regional politics (Cox 1993). Consequently, the power of the nation-state to control multinational money has declined. Investments are negotiated between international finance capital and the local powers try to do their best to attract firms. In this vein, public-private partnership and city marketing throughout cities and regions of the developed world have been emphasised (Kearns & Philo 1993).

Concurrently with the growing entrepreneurship of local and regional administration, the structural economic change is said to be facilitating the locally specific forms ('localisation') of economic activity (Swyngedouw 1992, 41). The rise of local economic agglomerations based on 'flexible specialisation, or, accumulation' and 'economies of scope' seems to offer new opportunities for territorial development. We even find hope for an entirely new, non-hierarchical spatial order reinforcing the specialisation and networking character of urban regions. This would also open new horizons for active city strategies, chiefly by intraregional and interregional networking (Camagni 1993; Cooke & Morgan 1993).

The shift towards a new economic-geographic order may, however, have been overemphasised. There are still only a few examples of strong industrial districts. Baden-Württemberg and the Third Italy are the ones usually mentioned outside of the traditional economic cores of Europe (Amin 1992; on an evolving Nordic discussion see Isaksen 1993). Localisation and globalisation, or glocalisation (Swyngedouw 1992), of the economy also create a paradox to the development of localities. While competing for investments, the entrepreneurial cities and regions may displace the investments in public infrastructure. Regions are not only independent actors in space but fundamentally *places* for people (Beynon & Hudson 1993, 182–183, Swyngedouw

1992). Questions that should also be asked are which powers take part in regional coalitions and who do they represent – people or corporate structures?

6.2.4 A Mosaic of Individuality or Inequality

The idea of a Europe of Regions is bound to the bottom-up strategy in regional development which places stress upon the individual nature of each 'region' with respect to the universality of capitalism and the nation state. In this perspective, regional development is based upon the idea of *diversity* rather than some unidimensional concepts such as growth or equality (Stöhr & Tödtling 1977). Asheim (1985) has argued that this perspective is ill-fitting with regard to the Nordic peripheries which depend very deeply on state intervention from the national level. Therefore, it is not certain that the new economic-geographic order, despite its increased freedom for local or regional action, would be more beneficial for the peripheral areas than any other strategy (Amin & Malmberg 1992). We must take seriously the warning of Beynon and Hudson (1993, 178): "For all the optimistic talk of 'A Europe of Regions', the future is likely to be one of growing disparities, of more losers than winners amongst the EC's towns, cities and regions as they strive to position themselves more favourably in relation to the processes of economic restructuring".

6.3 The Case of Nordic Peripheries

First, some basic peculiarities of local and regional governments in (three) Nordic countries must be noted. They all have a rather strong local government and a twofold statist/self-governmental system of regional administration. The main internal differences are related to the scope of regional self-government with Finland, on one hand, and Norway and Sweden, on the other hand. While the latter countries have an independent body of 'regional' self-government (*fylkeskommun/landsting*), the Finnish system is based solely on inter-municipal federations (*maakuntien liitot*)[6].

---

[6] All these three are called *regional councils* in the sequel.

In all of these countries the prime tasks of the regional bodies are confined to service provision. Their active development role is of later origin. In the Nordic countries regionalism, in the strict sense of the word, has never been a central part of national politics. The regional administrative level has been less extensive and less powerful than the local level. Therefore it is much more difficult to speak about wider institutionalised regions as actors in the European arena, compared to the situation in Southern and Central Europe.

In Nordic discussions on regional policy many writers emphasise the importance of a strong regional identity as a factor in regional development (e.g., Ekman 1991; Veggeland 1993). In response to the lack of active regionalism, regional development workers and politicians try to encourage spatial consciousness through the use of marketing campaigns, seminars, regional newspapers and influential persons channelling the 'voice of a region' (e.g., Paasi 1990; Östhol 1991). As Sogge and Imset (1993, 40) claim, there is still a long jump from identity regions to functional regions. The formation of regions is really not only a task for politics.

At the same time, there is a great demand for more competitive regions because of internationalisation. This has led to some recent reform plans in regional administration. In Sweden, for example, it has been proposed that the prevailing regional administration (24 state regional administrative units (län) / 25 regional councils (landsting)) be joined to form five or six territorially larger and more competitive units (SOU 1992:63). Similar arguments for larger administrative units both at municipal and regional levels have been put forth in Finland (Nummela & Ryynänen 1993) and in Norway (NOU 1992:15).

There has been, however, considerable disagreement on these plans. In the Nordic countries there is a long tradition of municipal autonomy. Municipalities form the most competent administrative unit below the nation state. Especially in Finland, regional councils are economically and politically dependent on both the municipalities and the nation state for finances and legislation.

Nevertheless, the prospects of Nordic local governance and its regional counterparts in regional development should not be underestimated in the European context. The rather long tradition of transborder co-operation in the Nordic countries is another important factor. There are several co-operative regional schemes across national borders, which have also formed a part of common regional policy among Nordic countries (see Nordic regions and transfrontier co-operation 1991) These co-operation schemes provide some challenges in evaluating the most idealist scenarios of a Europe of Regions (Sogge & Imset 1993, 40).

The creation of a regional planning system after the Second World War has helped establish the actor role of expanding local and regional administration in Nordic countries. In many countries, regional planning has created the institutional infrastructure, 'the regional machinery', that has been important in decentralising state affairs as well as enhancing population participation. Although planning was originally an extension of the central state's power, it has since become a tool for endogenous regional development. (Sharpe 1992, 12–13.)

In Finland, for example, a new regional policy idea has been launched which is based on the active role of the regional councils in formulating development programmes. However, to what extent these programmes should be truly 'regional', referring to wider regional administrative areas remains unclear. Some argue that it would be better to realize them at the local level corresponding to the travel-to-work areas (cf. the arguments of Hebbert 1993, 712, in the case of Britain).

According to the latter viewpoint it is the inter-municipal action at the local level which more directly meets the needs of functional regional system as well as boosting the chances of strong local administration in the Nordic countries. The functional urban regions surrounding the provincial centres have formed the main expanding frame of the community and settlement system of Nordic peripheral areas since the mid-1970s (Ventura & Wärneryd 1983; Vartiainen 1989). From this perspective, Nordic-style urban regions establish the main human resource basis for the northern peripheries in functional European networks. Still in these areas the social reality cannot be analysed exclusively through the logic of the urban society. The daily urban system is not far-reaching enough, either in a territorial or in a social meaning (Vartiainen 1989, 212; Carlsson et al. 1993).

In Sweden some major endeavours in regionalization have been directed towards wider 'Euroregions'. The Mälarregion around Stockholm and the West Sweden around Gothenburg represent ideas of large scale functional city regions. Skåne in the southern corner of Sweden is, in turn, a region that already has a strong historical identity which is exhibited via, for example, a special dialect. It is also a fundamental element of an evolving 'euroregion', comprising the West Baltic Sea

Region (Veggeland et al. 1993, 142), being the main land route from Scandinavia to Central Europe. The Bothnia Region in Northern Sweden and its evolving transnational links with neighbouring areas represents an attempt to create a strong northern region as an answer to the challenges coming from the south.

6.4 Conclusions

It is obvious that the main challenge for any individual region within a Europe of Regions is to strengthen itself through both intra-regional and inter-regional integration. This may be analysed strategically from four different angles (Figure 6.3).

The vertical-functional integration dimension of the Figure 6.3 illustrates the strengthening position in the global networks of the economy, while the horizontal-functional integration refers to transregional and transborder co-operation. The territorial integration refers to the strengthening of local networks within the economy as well as in cultural and political arenas.

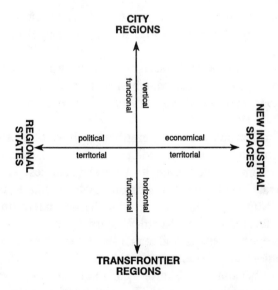

Figure 6.3. Four perspectives of the strengthening of a region

In developing the first strategy, the urban or nodal region, the goal is to develop a growth pole. According to Amin & Thrift (1992, 576) or Selstad (1992, 16), the centralisation of regional systems seemingly increases through sharper horizontal networks (Camagni 1993). Which agglomerations will succeed in this competition is difficult to predict. An implicit growth pole strategy has been a fundamental feature in the state policy of Nordic countries, where the regional centres have taken advantage of dispersed infrastructure investments in social and health policy and education. Consequently, their success has been founded on a strong public sector involvement in both employment and infrastructure which is now threatened by the public sector financial crisis. A call for more indigenous development efforts has been used as a basis for cutting public investments. However, it may result in greater inequality in regional development of the peripheries.

Alternatively, it may be that the future of the Northern peripheries is to serve as a vast green belt for Urban Europe. Veggeland et al. (1993, 97) presume that "the Nordic periphery, infrastructurally integrated, rich in resources and environmentally well preserved, may increasingly come to be regarded as a valuable source of recreational welfare and profitable tourism".

The second dimension of Figure 6.3, transborder co-operation, is widely believed to offer possibilities for the development of the peripheral localities. While the Nordic core areas are orientated primarily towards Western Europe and, occasionally, towards the eastern parts of the Baltic Sea Region, Nordic peripheries are searching for co-operation with each other and with their northern and eastern neighbours (cf. Veggeland et al. 1993, part V).

At the 'top of Europe', the very northern regions of Norway, Sweden, Finland and Russia, the Barents co-operation region, was established in 1993 with the aim of promoting its natural resources, such as timber, gas and minerals (Dellenbrant & Olsson 1994; Jussila et al. 1993). The future of the Nordic peripheries seems to be dependent upon the future of economic development in North-Western Russia. So, seen from the north, the European scene is wider than in the traditional discussion of a Europe of Regions (cf. Eskelinen 1994).

The third dimension of Figure 6.3 refers to the idea of locally agglomerated production systems. This is a contrast to the contemporary orientation of peripheries toward resource-based or labour-intensive phases in the externally determined production processes. Up to now, it has been an ideal which basically has been tied to the fourth angle: regions as autonomous political and cultural units. They both are connected to the bottom-up strategy in regional development which territorially corresponds to the ideal of Selstad (1993) and others on ecologically and culturally based regions (see also Vartiainen 1987).

The obstacles in implementing the territorial strategy are well-known in the contemporary logic of the globalizing economy. It seems that territorial integration is only possible by simultaneously deepening functional integration, i.e., by prompting industrial and political complexes within the context of expanding global corporate networks (Eskelinen 1984; Amin & Thrift 1992).

To conclude, we can offer two conflicting scenarios of the Europe of Regions from the standpoint of the Nordic peripheries. Firstly, a weakening national welfare system leaves the periphery to its own fate in a world of freely competing regions. Secondly, a self-reliance strategy promises lessening external dependencies in the international economy and more freedom from national or supranational political decision-making.

This tension can only be settled by a multitude of development efforts on different spatial scales and through different territorial formations. In adding new *sub*regional and *West*-European elements, we should not pass over the very Nordic tradition of regional thinking. While the traditional meaning of physical distance as such is loosing its relevance, the question of peripherality is even more connected with the situated (place-characterised) features of social reality. The future of the Nordic peripheries will depend upon whether they can realise a development strategy which is based on an interplay of national, local and Nordic scales both in the West and the East European context.

References

Amin, A. (1992), 'Big firms versus the regions in the Single European Market', In Dunford, M. & G. Kafkalas (eds.), *Cities and regions in the new Europe*, Belhaven Press, London.

Amin, A. & A. Malmberg (1992), 'Competing structural and institutional influences on the geography of production in Europe', *Environment and Planning A*, 24, 3, 401–416.

Amin, A. & N. Thrift (1992), 'Neo-Marshallian nodes in global networks', *International Journal of Urban and Regional Research*, 24, 4, 571–587.

Asheim, B. T. (1985), '"Top-down" -strategi: en nødvendig forutsetning for utvikling av periferien? Noen idéhistoriske, teoretiske og ideologikritiske refleksjoner', I Lomøy, J. (red.), *Geografi og planlegging*, NSGF Skrifter, nr. 13.

Beynon, H. & R. Hudson (1993), 'Place and space in contemporary Europe: some lessons and reflections', *Antipode*, 25, 3, 177-190.

Caciagli, M. (1990), 'Das Europa der Regionen. Regressive Utopie oder politische Perpektive?', *Österreichische Zeitschrift für Politikwissenschaft*, 4, 421-432.

Camagni, R. (1993), 'From city hierarchy to city network', In Lakshamanan T.R. & P. Nijkamp (eds.), *Structure and Change in the Space Economy*, Springer-Verlag, Berlin.

Carlsson, F., M. Johansson, L.O. Persson & B. Tegsjö (1993), *Creating Labour Market Areas and Employment Zones*, New Regional Division in Sweden Based on Commuting Statistics, CERUM report, Umeå.

Charpentier J. & C. Engel (eds.) (1992), *Les régions de l'espace communautaire*, Nancy.

Cooke, P. (1992), 'Regional innovation systems: competitive regulation in the new Europe', *Geoforum*, 23, 3, 365-38.

Cooke, P. & K. Morgan (1993), 'The network paradigm: new departures in corporate and regional development', *Society and Space*, 11, 5, 543-564.

Cox, K.R. (1993), 'The local and global in the new urban politics: a critical view', *Society and Space*, 11, 4, 433-448.

Dellenbrant, J. Å. & M-O. Olsson (eds.) (1994), *The Barents Region*, Security and Economic Development in the European North. CERUM, Umeå.

Dréze J. (1993), 'Regions of Europe', *Economic Policy*, October 1993, 265-287.

The European Parliament (1988), Resolution of 18 November, 1988. Annex OJ No. c326, 19.12.1988.

Ekman, A.-K. (1991), 'Bottom-Up strategies in the theory and practice', *NordREFO*, 4, 23-28.

Eskelinen, H. (1984), *On a reorientation of regional policies: Economic aspects of territorial approaches*, University of Joensuu, Publications of Karelian Institute No 68.

Eskelinen, H. (1994), 'Russian Karelia as a Peripheral Gateway Region', In Eskelinen H., J. Oksa, & D. Austin (eds.), *Russian Karelia in Search of a New Role*, Karelian Institute, University of Joensuu.

Goldsmith, M. (1992), 'Local government', *Urban Studies*, 29, 3/4, 393-410.

Goldsmith, M. (1993), 'The europeanisation of local government', *Urban Studies*, 30, 4/5, 683-699.

Harvey, D. (1989), 'From managerialism to entrepreneurialism: the transformation of urban governance in late capitalism', *Geografiska Annaler*, 71B, 1, 3-17.

Hebbert, M. (1989), 'Britain in a Europe of Regions', In Garside, P.L. & M. Hebbert (eds.), *British Regionalism 1900-2000*, Mansell Publishing Limited, London.

Hebbert, M. (1993), '1992: Myth and aftermyth', *Regional Studies*, 27 (8), 709-718.

Isaksen, A. (ed.) (1993), 'Spesialiserte produksjonsområder i Norden'. *Nordisk Samhällsgeografisk Tidskrift*, Uppsala.

Jussila, H., L.O. Persson & U. Wiberg (1993), *Shifts in Systems at the Top of Europe*, FORA, Stockholm.

Kearns, G. & C. Philo (eds.) (1993), *Selling Places. The City as Cultural Capital, Past and Present*, Pergamon Press, Oxford.

Keating, M. (1988), *State and Regional Nationalism. Territorial politics in the European state*, Harvester Wheatsheaf, Hemel Hempstead.

Keating, M. (1993), 'The continental meso: regions in the European Community', In Sharpe, L.J. (ed.), *The Rise of the Meso Government in Europe*, SAGE Modern politics series, Vol 32, London.

Kokkonen, M. & P. Vartiainen (1993), *Alueiden Eurooppa – haasteet alueelliselle kehittämiselle*, Sisäasiainministeriö, Kunta- ja aluekehitysosaston julkaisu 1, Helsinki.

Kunzmann, K.R. & M. Wegener (1991), *The pattern of urbanization in Western Europe 1960–1990*, Institut für Raumplanung, Universität Dortmund, Berichte 28.

Lever W.F. (1993), 'Competition within the European urban system', *Urban Studies* 30(6), 935–948.

Martin, S. (1993), 'The europeanisation of local authorities: Challenges for rural areas', *Journal of Rural Studies*, 9(2), 153–161.

Nijkamp, P. (1993), 'Towards a network of regions: The United States of Europe', *European Planning Studies*, 1(2), 149–168.

NOGRAN (1994), *Regionalpolitiska stödområden i Norden*, Snabbrapport No 2, Helsingfors.

Nordic Regions and Transfrontier Cooperation (1991), Nordic Group for Regional Analyses, The Secreteriat of the Nordic Council of Ministers, Borgå.

NOU 1992:15, *Kommune- og fylkesinndelingen i et Norge i forandring*, Oslo.

Nummela, J. & A. Ryynänen (1993), *Kuntahallinnon vaihtoehdot*, Lakimiesliiton Kustannus, Helsinki.

Paasi, A. (1991), 'Regional identitet: ett handelsvara eller en mental förutsättning för utveckling?', *Nordrevy*, 1, 13–19.

Regional utveckling i Norden (1993), Basprojektets årsrapport 1993/94, *Nord*, Vol 30, Århus.

Rokkan, S. & D.W. Urwin (1983), *The Politics of Territorial Identity*, Sage, Beverly Hills.

Selstad, T. (1992), 'Agglomerer eller dø!', *Nordrevy*, 6, 13–18.

Selstad, T. (1993), 'Nordens natur – periferiens framtid?', *Nordrevy*, 3/4, 50–60.

Sharpe, L. (ed.) (1992), *The Rise of Meso Government in Europe*, SAGE Modern politics series, Vol 32, London.

Sogge, S. & Ø. Imset (1993), 'Bananer i lange baner', *Nordrevy*, 5, 39–41.

SOU 1992:63, *Regionala roller. En perspektivstudie. Betänkande av regionutredningen*, Stockholm.

Stoker, G. (1991), 'Trends in Western European local government', In Batley, R. & G. Stoker (eds.), *Local Government in Europe*, Macmillan, Basingstoke.

Stöhr, W. & F. Tödtling (1977). 'Spatial equity – some anti-theses to current regional development strategy', *Papers of the Regional Science Association*, vol. 38, 33–53.

Swyngedouw, E. (1992), 'The Mammon quest. 'Glocalisation', interspatial competition and the monetary order: The construction of new scales', In Dunford, M. & G. Kafkalas (eds.), *Cities and Regions in the New Europe*, Belhaven Press, London.

Vartiainen, P. (1987), 'The strategy of territorial integration in regional development: Defining territoriality', *Geoforum*, 18(1), 117–126.

Vartiainen, P. (1989), 'Counterurbanisation: a challenge for socio-theoretical geography', *Journal of Rural Studies*, 5(3), 217–225.

Veggeland, N. et al. (1993), *Impact of the Development of the Nordic Countries on Regional Development and Spatial Organisation in the Community*, The Synthesis Report. Nordic Institute of Regional Policy Research. Copenhagen.

Ventura, F. & O. Wärneryd (1983), 'Differentiation of settlement systems on the basis of population densities and level of development. 'Regionalization' now and in the future', *Geographia Polonica*, 47, 29–37.

Östhol, A. (1991), 'Nyregionalism – modell för Sverige?', *Plan*, 45(5–6), 251–255.

CHAPTER 7

Peripherality and European Integration: the Challenge Facing the Nordic Countries

Sven Illeris
Roskilde University

7.1 Introduction

The purpose of this chapter is to discuss the consequences of the peripherality of the Nordic countries, in the context of European integration. We shall first try to clarify these concepts, in particular peripherality, and highlight the situation of the Nordic countries relative to other parts of Europe at the present time and previously. We shall then address the question of *where* the consequences of integration between the Nordic countries and the EU will occur. Finally, we shall discuss the question of *what* the consequences are likely to be.

By the *Nordic countries,* we mean Denmark, Finland, Iceland, Norway, and Sweden. These countries are closely linked, culturally and economically. The populations also largely share the same attitudes and have constructed similar welfare states. Politically, the countries have never committed themselves to any close mutual co-operation, although it should be mentioned that they have agreed to admit all Nordic citizens into their labour markets. Denmark has been a member of the European Community since 1973 and will in some respects be treated together with the other EU countries, while Finland and Sweden have joined the EU in 1995, after this chapter was written.

European integration should be understood as something far broader than EU membership and the precise measures of integration deriving from it. We might have used the words internationalisation or globalisation, but most of the interaction of the Nordic countries is with other European countries. The term integration will include trade in goods and services, capital flows, foreign direct investment and ownership by transnational companies, networks and strategic alliances, and also interaction which is not economic in a narrow sense, such as migration, information and innovation flows, as well as scientific, artistic and cultural impulses.

All of these flows grow rapidly, at rates far above GDP growth. Even if a country were to isolate itself from these flows and influences, as Albania did, it is no longer possible. If a country chooses not to be a member of the EU, it will still have to follow the rates of interest of the Bundesbank and the technical standards set by the EU – it only cuts itself off from having an influence on the rules of the game. If its firms are not owned by transnationals, their products will still have to compete with the products of the latter on domestic and foreign markets. Hence its firms must have conditions of production, factor prices and regulatory frameworks which do not radically differ from those of their competitors. Sovereignty is largely an illusion, membership in the EU or not does not make any great difference.

7.2 Peripherality, Accessibility, and Development

Peripherality, the key concept of this chapter, is often used in a rather imprecise way. One may distinguish between the following characteristics:

In itself, the word signifies a peripheral situation in relation to a centre or core. In Europe, this core has for four centuries been in the regions surrounding the English Channel. The term is, however, purely geometric and does not say whether this situation is good or bad.

An inevitable consequence of peripherality is low accessibility, compared to the core. The cost of interaction with other regions is high because of longer distances.

However, the differences in accessibility have been reduced dramatically in the course of history, due to improved and cheaper means of transport and telecommunications. Figure 7.1 shows how the creation of airlines has reduced the loss of time – often the heaviest transport cost. Travelling from one end to the other of a small country took 8 hours in 1960 and less than 3 hours (city centre to city centre) in 1986, when the network of domestic airlines had been established. This process still advances, as witnessed by the data of Erlandsson (1991, 1993) on accessibility, measured as the number of persons in other European cities who can make a 4-hour visit to the city in question in one day. For cities in the core of Western Europe, accessibility increased by 26 per cent from 1976 to 1988, and by 5 per cent from 1988 to 1992. But for cities in the periphery, accessibility increased by 29 per cent and 12 per cent, respectively. Thus, if the peripherality of the Nordic countries is still a handicap in the sense of lower accessibility than in the core regions, it is a handicap which has been reduced by now.

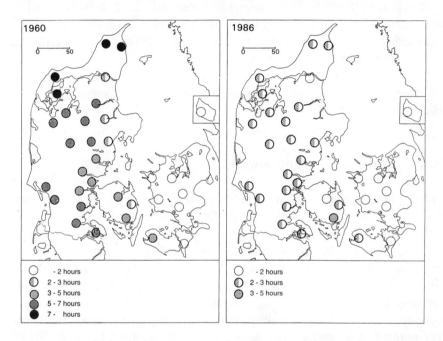

Figure 7.1. Fastest travel by public transport between Copenhagen and provincial Danish towns, winter weekdays 1960 and 1986.

Low accessibility may lead to a low level of trade, fewer cultural impulses etc., ending in a state of relative isolation. This is a severe problem. Porter (1990) argues that isolation, the lack of competitive challenges and innovative impulses, leads to inbreeding and stagnation. This point of view can be extended from the economic to the cultural sphere. Exposure and division of labour, on the other hand, will have long-term benefits for innovation and dynamism which are far more important than the tiny short-term cost reductions mapped by the Cecchini study (1988).

Other people have had different opinions. The "self-reliance" strategy implies at least a strong reduction of the interaction with the rest of the world – especially with the most advanced regions. The former leaders of Albania consciously did their best to isolate the country. We know the results.

However, low accessibility does not necessarily lead to isolation: The actors may be able and willing to pay the costs of good connections with other regions. The number of Swedish tourists on Mediterranean beaches has been large for years.

Isolation is not necessarily a companion of peripherality, but undoubtedly some economic protectionism and cultural isolationism may be found in the Nordic countries.

7.3 Peripherality and low Density

Other characteristics which sometimes have been connected with the concept of peripherality are not directly derived from distance and low accessibility.

Thus, peripherality is sometimes connected with low densities of population and economic activities. It is true that cores always display high densities and include big cities. But peripheries are not necessarily sparsely populated. Most of Mediterranean Europe is not. The Nordic countries do have very low densities: If Denmark is excluded, the average number of inhabitants per square km is 15, against 145 in the EU of 12.

As discussed by Mønnesland in this book (see ch. 8), low densities may form a handicap in several respects. In the membership negotiations with the Nordic countries which were concluded in 1994, the EU has recognised the handicaps of low densities by defining a new objective (number 6) of its structural funds, namely to promote development in areas with less than 8 persons per square kilometre.

7.4 Peripherality and Poverty

Peripherality has often been connected with economic backwardness, or more precisely with low GDP per capita. While it is true that the Mediterranean and Atlantic peripheries of Western Europe are poor, compared to the core, this is on the whole not the case in the Nordic countries, see Table 7.1.

Table 7.1. GDP per capita 1990, by Purchasing Power Parities

Index EC 12 = 100			
Belgium	107	Denmark	111
Germany	118	Finland	113
Greece	55	Iceland	122
Spain	82	Norway	127
France	113	Sweden	117
Ireland	72		
Italy	107		
Luxembourg	136		
Netherlands	107		
Portugal	58		
UK	110		

Source: Foss et al. (1993)

When GDP levels are compared using exchange rates, for many years the Nordic countries have been registered among the world's richest. This has been exaggerated, however, and the devalued Nordic currencies now better reflect their purchasing power. Even so, they are prosperous countries. Even their northern regions show high incomes, but the incomes are inflated by transfer payments and agricultural subsidies.

Of course, the measurement of GDP is subject to heavy criticism, especially in the Nordic countries where the public sector, without marketed outputs, is substantial. However, if other indicators of economic level are used, as in Table 7.2, the results remain that the Nordic countries have a high economic quality of life.

Table 7.2. Indicators of quality of life

	Share of population aged 10–64, 1989 with highest educational attainment:		Telephone main lines per 1000 inhabitants 1990
	Primery school	University (3 years)	
Belgium	63	7	401
Germany (W)	22	10	432
Spain	80	9	324
France	50	7	482
Ireland	62	7	279
Italy	74	6	387
Netherlands	45	6	462
Portugal	93	4	226
UK	35	9	444
Denmark	43	10	566
Finland	42	10	534
Norway	35	11	502
Sweden	33	12	681

Sources: Illeris & Sjøholt: Producer Services (1993), Yearbook of Common Carrier Telecommunication Statistics

7.5 Peripherality, Exploitation, and Dependence

Finally, the concept of peripherality is often connected with exploitation from the core. Exploitation may operate via various mechanisms. We shall focus here on two examples:

One is to reserve the high-value-adding activities for the core, and leave only low-productivity, low-wage activities to the periphery – typically the production of raw materials from local natural resources.

The Nordic economies are indeed largely based on natural resources: Agriculture in Denmark, fisheries in Norway, Iceland and Denmark, forestry in Sweden, Finland and Norway, oil and natural gas in Norway and Denmark, hydroelectric power in Norway, Sweden, Finland and

Iceland, iron and other ores in Sweden, Finland and Norway (see examples in Table 7.3).

Table 7.3. Examples of primary production

	Fish catches 1990 1000 t	Roundwood 1990 mill. m³	Hydro-electricity 1990 1000 GWh	Crude oil 1991 mill. t	Iron ore 1991 1000 t
Belgium	42	4.7	1	0	0
Germany (W)	250	73.5	18	3.4	n.a.
Greece	140	2.0	2	0.8	600
Spain	1458	17.8	26	1.1	1604
France	897	44.7	57	3.0	2256
Ireland	231	1.5	1	0	0
Italy	525	8.0	35	4.3	0
Netherlands	438	1.4	0	3.3	0
Portugal	322	10.4	n.a.	n.a.	4
UK	804	6.5	7	86.8	2
Denmark(1)	1517	2.1	0	7.1	0
Finland	97	41.6	11	0	33
Iceland	1508	0	4	0	0
Norway	1747	11.8	121	91.2	1436
Sweden	260	55.9	73	0	13046

(1) Excluding the Faroe Islands and Greenland, where 283,000 tons and 138,000 tons of fish were caught
Sources: UN Statistical Yearbook, International Energy Agency: Energy Statistics of OECD Countries, UN Yearbook of Industrial Statistics

Some of these raw materials are exported with little value added. For instance, the Norwegians do not refine much of their oil and fish. But on the whole, the extraction itself has high productivity and uses refined technology. In some cases Nordic firms are world leaders. Also, many raw materials are processed, moving them well towards final products, as witnessed by the paper industry in Finland and Sweden, by the aluminium smelting based on hydro-electric power in Norway and Iceland, and by the Swedish engineering industry. One cannot interpret the natural resource base of the Nordic economies as evidence of exploitation by the core.

Another mechanism of exploitation which should be discussed is core ownership of firms operating in the periphery. This would mean that the core made the periphery dependent, controlled it and extracted profits from it.

In this respect, the Nordic countries actually show the opposite picture. Nordic outward direct investments, mainly in the UK, Germany, France and the Netherlands, are far greater than the inward investments. Indeed, Swedish, Norwegian and Finnish direct investment in the EC around 1990 – caused partly by the fear of being excluded from the Single Market – undermined domestic investment and is part of the explanation of the present economic crisis of these countries. It is rather too little foreign investment – with the accompanying knowledge and innovation – than too much that is the Nordic problem.

To conclude, the Nordic countries are peripheral in a geometric sense and hence have relatively low accessibilities. They have also problems caused by the low densities of population. But they have none of the other problems sometimes attributed to peripherality, such as low GDP per capita, low quality of life, exploitation and dependence. However, this has not always been the case.

7.6 Nordic Peripherality Throughout History

As far back as economic estimates can be made, the Nordic countries have formed not only a peripheral, but also a poor and backward part of Europe. This was the case in Antiquity, when Greece and later the Roman Empire was the core region. It was also the case in the Middle Ages when Northern Italy formed the core, and in the Modern Epoch when the centre moved to the countries surrounding the English Channel. In the 50 years preceding World War I, Sweden and Norway were so poor that 1.5 million people emigrated to other continents, mainly America, corresponding to 20 per cent of the population of these two countries in the year 1900.

Two main reasons explain this relative backwardness: First, before industrialisation natural resources, especially agricultural resources, were very important for the production and standard of living of a region. Also, the soils and climates of the Nordic countries were poor, with the exception of Denmark. Second, when transport was extremely expensive and slow, and telecommunications hardly existed, low accessibility was a far more severe handicap than today. Innovations from the core found their way only very slowly into these remote outskirts of Europe.

The Nordic countries were industrialised relatively late, compared to the European core. Why have they then become some of the most prosperous European countries during the last century? It is difficult to define the determinants of long-term economic growth with respect to countries and regions. Perhaps the early widespread literacy which the Lutheran churches promoted in the Nordic countries is partly responsible. Adult education stressing bottom-up and independent enlightenment, which Grundtvig advocated here, has probably contributed as well. The strong Social-Democratic governments which were able to maintain an optimal balance between exposure and protection allowed the egalitarian welfare state to redistribute resources and demand to all groups of the population (Senghaas 1982).

7.7 Where Will European Integration Have an Impact?

We shall now turn to the question of where the consequences – positive or negative – of closer integration between the Nordic countries and the EU of 12 will occur. The impact of increased integration between regions is assumed to be larger, the higher the amount of interaction between regions. If there is little interaction – as between northern Finland and northern Greece – increased integration is not likely to have much impact on either region. It would have been useful to discuss interaction between regions, as done by Cappellin in this book (see ch. 3), but the current approach requires data on interaction which are only available between countries.

The best data available on interaction refer to international trade in goods. Such data on the trade between the EU12 member countries and the 4 Nordic countries are shown in Table 7.4.

Table 7.4. Trade between EC countries and four Nordic countries, 1990

Per cent of total international trade	Exports to Nordic countries	Imports from Nordic countries	Total to/from Nordic countries
Belgium-Luxembourg	2.5	3.7	3.1
Denmark	20.8	19.5	20.2
Germany (W)	4.6	4.9	4.8
Greece	2.8	2.7	2.7
Spain	2.0	3.0	2.6
France	2.2	3.7	3.0
Ireland	3.3	3.2	3.2
Italy	2.3	2.5	2.4
Netherlands	3.0	4.6	3.8
Portugal	7.1	3.6	5.0
UK	4.9	7.8	6.5
	exports to EU12	imports from EU12	total to/from EU12
Finland	45.1	45.8	45.5
Iceland	67.7	49.7	58.6
Norway	64.9	46.3	56.7
Sweden	54.0	54.8	54.4

Source: UN International Trade Statistics Yearbook

For Denmark, the other Nordic countries are important trading partners, but for other EU countries, they are only partners of secondary importance, accounting for 4–7 per cent of the total international trade of the UK, Germany, the Netherlands and Portugal, and for 2–3 per cent in the rest of the Community.

For the Nordic countries, on the other hand, the EU accounts for more than half of their total international trade (for Finland in 1990 a little less). The impact of integration is likely to be considerable.

In the EU, on the other hand, it is the northern EU countries, where distances to the Nordic countries are short and transport costs low, that have substantial trade with the Nordic countries and can expect substantial consequences of integration.

But can distances, and of course the size of the economies, explain the pattern of trade and other interactions? Inspired by Haass & Peschel (1982) we have used an approach not often applied in international trade research, namely the so-called gravity model. In its crudest version, it states that:

$$I_{ij} = K \frac{m_i m_j}{d_{ij}^a} \quad \text{in which}$$

- I_{ij} is the amount of interaction between the regions i and j,
- m_i and m_j are the "masses" of the regions i and j, measured for instance by the GDP of these regions or by their population,
- d_{ij} is the distance or transport costs between the regions i and j, measured for instance by kilometres, time distance, or transport costs.
- a is an exponent which shows whether the interaction decreases slowly or rapidly when d_{ij} increases.

This model has been shown to fit well to many observed types of interaction, for instance migration, traffic, telephone calls and retail shopping. It does not offer a theoretical explanation of causal relationships, but is a good descriptive tool.

The analysis will consist of using an iterative procedure to fit this model to the observed matrices of interaction between West European countries, using various measures of "masses" and distances. In this way we can:
- determine whether the model gives a good fit (as expressed by correlation coefficients); in other words whether the observed interaction does increase when "masses" increase and distances decrease,
- determine which measures of "mass" and distance give the best fit,
- determine the size of the distance exponent a that gives the best fit, in other words how steeply interaction decreases by increasing distance, and
- when the model has been calibrated with the best fitting measures of masses and distances, calculate the expected amount of interaction

between each pair of regions, and then see how the observed interaction between the regions deviates positively or negatively from the amount estimated by the model. The deviations or residuals indicate the influence of other factors than masses and distances, for instance cultural barriers and affinities.

The model has been applied to international trade in 1990. Air distances "as the crow flies" between the centres of gravity of the population of West European countries are used. The measure of "mass" is the 1989 GDP (a first try with population as mass did not give a good fit). Three versions have been applied. In the first version, only the nine small, "rich" countries were included (the five Nordic countries, the Netherlands, Belgium-Luxembourg, Switzerland, and Austria). In the second version, the three small, "poor" countries of Ireland, Portugal and Greece were added. In the third version, also the five big countries of W. Germany, the UK, France, Italy, and Spain were included. Table 7.5 shows the correlation coefficients and distance exponents obtained.

Table 7.5. Results of gravity model analysis of international trade 1990

	Correlation coefficient r^2	Distance exponent a
Version 1	0.905	1.527
Version 2	0.932	1.244
Version 3	0.857	1.487

The high correlation coefficients show that the model does fit the observed pattern of trade well. The distance exponents are not surprising. We shall not discuss them in detail.

On the deviations between observed and estimated interaction, the following remarks can be made:

The five big countries show (in version 3) several negative deviations, especially in trade with other big countries. This is not surprising. Because of the size of these countries, economic actors are more likely than actors in the small countries to find domestic trading partners. The opposite is the case in the smallest country, Iceland (but the positive deviations of this country may also show that the friction of distance over the ocean is exaggerated by the air distance applied).

Positive deviations are observed in the mutual trade between the Nordic countries, to a more modest degree between the Benelux countries, between Germany, Switzerland and Austria, and between the UK and Ireland. These deviations must be interpreted as cultural affinities. It is – as many studies have shown in the Nordic countries – easier to buy and sell in other countries whose language, consumer preferences and general ways of thinking and trading are similar to one's own, than in culturally different countries – even if the distance to the latter is shorter (as from Denmark to Germany). Other positive deviations have various explanations, for instance the Icelandic export (of fish) to the Mediterranean countries and Switzerland.

The Mediterranean countries (except Portugal) have relatively little mutual trade – Greece, in particular, has little export of goods. However, this is to some degree compensated by tourism (export of services). Switzerland and Austria show negative deviations in almost all directions (except mutually and with Germany), which is more difficult to explain.

Johansson (1993) quotes a similar analysis for the years 1965-70-75. It shows strong positive deviations for trade between the Nordic countries, and positive deviations for trade between EU countries, between English-speaking countries, and between German-speaking countries.

Another set of data concern telecommunications using the telephone network. These data are interesting because they reflect all types of interaction: trade, administration, scientific, cultural, social and personal contacts etc. – they all use the vehicle of telecommunications.

Rossera (forthcoming) has analysed telephone contacts between all European and Mediterranean countries in 1986. He first grouped the 37 countries into 18 clusters in such a way that international telephone traffic was maximum within clusters and minimum between clusters. In Western Europe for instance, Finland, Norway and Sweden formed one cluster, West Germany and Austria another, the Iberian peninsula a third one, the Benelux countries a fourth one, and the British Isles a fifth one.

His next step was to analyse telephone traffic between the clusters, applying the gravity model in the same way as described above. As "masses", the total amount of outgoing and incoming traffic was used, and as distances kilometres between the main telecommunications centres of the clusters. On the whole, Western Europe turned out to be fairly evenly integrated, though a closer inspection of the results show that all strong negative deviations are between Germanic and Latin clusters, while most strong positive deviations are between Germanic clusters mutually or between Latin (including Greek) clusters mutually. There was also a positive deviation between the British Isles and the Iberian peninsula, probably connected with tourism. Again, Finland-Norway-Sweden, Denmark and Iceland are strongly interconnected. The Nordic countries seem to be rather isolated from the rest of Europe, but this is due to the fact that "masses" are measured as total outgoing and incoming traffic and hence "hypertrophied" because of the high volume of traffic between the Nordic countries mutually. In a series of regression analyses, Rossera examined factors which could explain positive deviations and found for instance that EU membership and former colonial ties were significant.

The conclusion of the analyses is confirmation of the notion that trade and other types of linkages are influenced by distance, and that hence, increased integration between the Nordic countries and the EU of 12 will primarily create impacts in the Nordic countries and the northern EU countries.

The analyses also showed cultural affinities to be important. The Nordic countries are very strongly mutually interconnected. This is confirmed by Törnqvist (1993) who quotes an assessment of "mental distances", made by Swedish company managers. They consider "mental distances" to other Nordic countries small, followed by "mental distances" to English- and German-speaking European and North American countries (including the Netherlands), while "mental distances" to Latin European countries are longer and to the rest of the World maximum.

Thus the Nordic countries form a "Randkerne", a secondary core in the periphery. According to Haass & Peschel (1982) they have done so for at least a century.

7.8 What Will the Regional Integration Impacts be?

As already asserted, increased economic and cultural interaction will be to the benefit of all societies involved. However, some actors and hence some regions will benefit more than others. Reduced barriers and more competition means that competitive firms will increase their sales at the cost of less competitive firms. But nothing can be said a priori about which firms and which regions are more competitive, and which are less. That must be examined in each particular case.

It is not possible in this chapter systematically to study all sectors, in order to find out whether Nordic firms are more or less competitive that their competitors in the Community, nor where increased integration is likely to result in the gain or loss of markets, turnover and employment. We shall, however, briefly survey some sectors.

As regards the production of goods, it has already been mentioned that the Nordic countries are strong producers and exporters of a number of raw materials and manufactured products based upon them. On the other hand, the Nordic countries are weak in agriculture and food industry and must expect substantial reductions from integration. Exceptions are the Danish textile and chemical industries. This is, however, a very crude picture, and there are numerous other exceptions both in the Nordic countries and the EU, which give witness to strong niches within otherwise weak sectors.

International competition is keenest in the goods markets. Services are underrepresented in international trade, with the exception of transport and tourism, and must be expected to remain so. Most household services depend on the frequent visits of customers to the provider, so large-scale international trade is impossible, e.g. in retailing and schools for small children. Producer services, too, often depend on the users' co-operation in the creation of the service, e.g. in consultancy, and hence on frequent face-to-face contacts. Even if the transport costs of meetings can be paid, good communication still requires cultural understanding. Therefore, exports remain relatively modest. More often, the exporter sets up a branch office in the market country, acquires an existing firm, or forms a strategic alliance, in all cases largely staffed by domestic labour.

Generally, the large cities in the core of Europe are strong in services. Accounting, advertising and management consultancy in the Nordic countries is dominated by mainly Anglo-Saxon multinationals. But for the above-mentioned reasons, it is not likely that they will oust many of the activities of the Nordic service-based cities. The Nordic countries also receive fewer tourists than they send abroad (with the exception of Denmark).

On the other hand, there are also Nordic strongholds in the service sectors, such as the Norwegian merchant fleet (the largest of any European country, with 24 million GRT in 1991) and the derived activities of classification and insurance of ships. Other examples are the Danish environmental consulting firms and software production (Illeris & Sjøholt, 1995).

There will be sectors and regions in the Nordic countries as well as in the EU which will lose from increased European integration, while other sectors and regions will gain. But for whole countries, there is no reason to expect dramatic consequences. The geometric peripherality of the Nordic countries does not condemn them to be losers. And in the long run, Europe as a whole will gain, both economically and culturally.

References

Cecchini, P. (1988), *The European Challenge 1992: The Benefits of a Single European Market*, Wildwood House, Aldershot.
Erlandsson, U. (1991), *Kontakt- och resemöjligheter i Europa 1976 och 1988*, Rapport 88, Institutionen för Kulturgeografi och Ekonomisk Geografi, Lund.
Erlandsson, U. & Lindell, C. (1993), *Svenska regioners kontakt- och resemöjligheter i Europa 1992*, Rapport 119. Institutionen för Kulturgeografi och Ekonomisk Geografi, Lund.
Foss, O. et al, (1993), Peripherality. In *Impact of the Development of the Nordic Countries on Regional Development and Spatial Organisation in the Community, vol. II*, NordREFO, Copenhagen.
Haass, J.M. & Peschel, K. (1982), *Räumliche Strukturen im internationalen Handel: eine Analyse der Aussenhandelsverflechtung westeuropäischer und nordamerikanischer Länder 1900-1977*, V. Florentz & Institut für Regionalforschung der Universität Kiel, München.
Illeris, S. & Sjøholt, P. (1994), *The Regional Distribution of Producer Services in the EC and the Nordic Countries*. Research Report no. 90, Department of Geography and International Development Studies, Roskilde University.
Illeris, S. & Sjöholt, P. (1995), 'The Nordic Countries: High Quality Service in a Low Density Environment', Chapter 8 in F. Moulaert & F. Tödtling (eds.), Special Issue 'The European Geography of Advanced Producer Service Firms', *Progress in Planning*.
Johansson, B. (1993), *Ekonomisk dynamik i Europa*, Liber-Hermods, Malmö.
Porter, M. (1990), *The Competitive Advantage of Nations*, Macmillan, London.
Rossera, F. (forthcoming), 'Communication Flows and Barriers to Communication in Europe', *Sistemi Urbani*.
Senghaas, D. (1982), *Von Europa lernen*, Suhrkamp, Frankfurt a.M.
Törnqvist, G. (1993), *Sverige i nätverkens Europa*, Liber-Hermods, Malmö.

CHAPTER 8

Regional Development in the Nordic Periphery

Jan Mønnesland
Norwegian Institute for Urban and Regional Research

8.1 Introduction

In Western European countries, the recent decade has been characterised by enforced economic integration. Within the EU, this process has been followed by a programme of a more federalistic type, denoted by the new term European Union (where the label Europe is used in the narrow-minded way rather common in the EU, i.e. not including the nearly three quarters of Europe not within the EU area). The EEA agreement between the EU and EFTA countries (except Switzerland) has created a juridical base for economic integration, aimed at reducing the hampering effect of national borders on market transactions. The recent enlargement of the EU, in which Austria, Sweden and Finland have enrolled as new EU members from the beginning of 1995, also adds to the integration process.

The integration process reduces the effect of national borders on general regional development. The effects of differences between regions, unaffected by the intra-EU (or intra-EEA) borders, are often expected to be more important in the spatial processes of Western Europe. The label "regionalization" is often used as a description of this new phenomenon, in the same way as the label "Europe of regions" has been launched as a successor to the earlier "Europe of nations".

Such labels reflect political intentions and interests. As usual, the scientific milieu will also be involved in the general struggle for creating the most handsome vocabulary. The central parts of the EU structure will have an interest in enforcing regional strength relative to the national level, as this will ease the way towards a federal structure according to the aims set up in the Maastricht Treaty. The nation states may see such a development as problematic, however. The current trend now seems to be in favour of national power versus federal in the political development of EU. It remains to be seen if this is only a temporary reaction to the somewhat premature Maastricht treaty perspectives, or if it will influence the integration process in a more substantial way.

As Western Europe is a rather heterogeneous territory, the general concepts and trends used to describe the ongoing process in particular areas, become often misleading when used to describe a process in other parts of the territory. On a Western European scale, the term "region" relative to "nation" may reflect different realities both in a cultural and economical context, making an unreflected use of general labels more misleading than illuminating.

The scope of this chapter will be to focus on the development in Norway, Sweden and Finland. The general processes of regional redistribution will be judged in the light of the special characteristics of these three Nordic countries, being small populations within wide areas, with long distances, both domestic and in relation to the core West European markets. The focus will be on the relations between centrally and peripherally located regions. The way redistributive forces operate in low density populated regions is developed both from the production and population side. The policy effect on such processes will also be assessed.

8.2 Regional Processes under Enforced Integration

In the last decades technological development has been rather rapid. The production structure has undergone important changes, where productivity growth has reduced the demand for labour in traditional industries. As the new industries have different spatial patterns than the old ones, this in itself is a factor generating regional redistribution. The technological development also creates better communications, both for goods, people and information. Therefore, competition between regions sharpens as shelters, created by geographical distance, lose their importance, the result often tending toward a polarisation between winner and loser regions.

The integration process occurring within the EU and EEA nations aims to remove formal and informal trade limitations on a rather large scale. The new GATT treaty, when effectively enforced, will also try to liberalise trade on the global level. Although the tendency on the global level is difficult to ascertain, the tendency towards greater integration within Western Europe, including the Nordic countries, seems clear as far as the legal framework is concerned.

Both the ongoing technological process and the West European integration process will create new conditions. The interregional competition will take place on an West European level as well as on the national level, and also generate stronger competition between nations. In many aspects, the structure of the international competition is parallel to the structure of interregional competition.

There are, however, important differences between interregional and international integration effects. These differences are both political and cultural.

Much attention has been directed towards national political actions which aim to prevent competition from abroad and to promote domestic export. Such actions tend to make interregional competition within nations stronger than competition between nations or between regions in different nation states, as the political efforts will serve as trade barriers. Such efforts are still strong in the global context. Also within the EEA and EU area, the level of non-tariff barriers is still of importance, but the EU actions have succeeded in reducing the effect of this type of policy.

One could expect intra-European competition to develop more or less in the same way as interregional competition within nations. The concept "Europe of regions" is partly based on such an argument, where the integration tendency is interpreted to include the dismantling of national barriers so that the regions will be the real units. Europe would then be the new nation state with common internal rules and common external trade barriers. In all such statements, the term "Europe" has to be interpreted in the narrow West European way. The rather effective trade barriers established by the EU towards non-EEA countries make such statements irrelevant in an all-European interpretation. The emerging of trade treaties between EU and several Central European countries has so far only modified and not cancelled the effect of the EU external trade barriers.

There are, however, important political action groups within the nation states, trying to influence the relations between domestic regions. For the Nordic countries, the egalitarian philosophy has resulted in a rather complex battery of regional policy actions which are intended to support the weaker and peripheral regions. Thus, there is not a pure laissez-faire competition between regions within the states, either in the

Nordic countries or in the continental EU countries. On the EU federative level as well, similar actions are taking place in order to limit growing regional imbalances.

Cultural barriers may be longer-lasting and sometimes even more powerful than political barriers. Both language and the legal style differ among the West European states, and partly also within the states. Such barriers hamper real integration, and may also create a base for restrictive attitudes towards further integration. In addition, pure economic reasons may also lead nations, regions, social segments and industries to find themselves better off keeping trade barriers, unaffected by the potential gains and losses of integration on the federal level.

In some cases, the cultural barriers between regions may also be rather strong within individual EU states. There may also be regions where actors believe their economic interests are more threatened by national actions than by EU economic integration. In such cases, a weakening of the nation state in order to create a Europe of regions may obtain strong support for cultural and/or economic reasons. Also, the national borders may not generally be important borderlines in a cultural context. Therefore, the cultural identity would not need to be challenged by enforced European integration. In a similar way, social segments which expect that they would be winners in the economic competition of an integrated internal market, will also tend to be in favour of integration.

The Nordic countries are mostly characterised by a high degree of national-cultural homogeneity. Of course there is also a high degree of regional patriotism. On the North Cap, the Sami people have an ethnicity quite different from the other Nordic nationalities. Even so, compared with the other European countries, the national homogeneity is high. The nation state fits well into the cultural feeling, creating a strong national loyalty. The situation observed in several other European regions, where regional patriotism leads to strong support for EU integration in order to override dominance from the capital region, is absent in the Nordic area. Compared to other non-Nordic European countries, the Nordic countries have a strong similarity between the national and the cultural borders.

The situation observed in several continental European countries, with several cases of cultural regional tension (the north-south divide in Italy, separatist tendencies within regions in Spain and the United Kingdom) together with unclear cultural divides along national borders in other areas (as around the Be-Ne-Lux), is not transferable to the Nordic area. There are no such tensions or cultural feelings calling for a reduced importance of the national borders. Therefore, the EU integration will have to be judged from a more rational base, where economic gains are weighed against the costs.

This does not imply that regional cross-border contacts are non-existent and that all communications are channelled via the national level. Contact between neighbouring regions have been more or less active in different historical phases. Such communication lines are then evidently not a new feature. It would be a mistake to regard such communications as a new trend, and there is no sign that such regional contacts are weakening the homogeneity and the cultural and political strength of the Nordic nations.

The claim that continued integration within EU creates a process of regionalization due partly to a weakening of the national level and partly to a border-free trade and competition pattern, seems to have no relevance for the Nordic countries so far. Also, there are no indications that such a process will be more profound in the future.

The extent to which the concept of regionalization can illuminate the process occurring within the continental EU countries, is not discussed further in this article. For the Nordic countries at least, the regional challenges created by West European integration will have to be analysed in the same way as has been the tradition in domestic regional studies. The scope, however, needs to be widened, to grasp the fact that nearly all the Nordic countries (except Denmark and the southernmost part of Sweden) are in the European periphery. Then, the national centres must also be analysed as regions with a peripheral location in the European market place.

8.3 Contrasting Continental Europe

The population density in continental Europe is usually more than ten times higher than in the Nordic countries (excl. Denmark), see Figure 8.1. This fact has important consequences for the regional structure, as well as for the regional effects of structural change and market integration. The distances from the northernmost periphery to the nearest town or urban agglomeration with more than 100,000 inhabitants are on a level unknown in the rest of Europe except the northern

Russian tundra. Travelling to "town", for many settlements, means journey-times up to half a day and more, and air travel over long distances must also be used. Even trips to the nearest urban agglomeration often need longer travel-time than one would need in more central parts of Europe to go from one capital to the next.

The location of Denmark makes it a non-typical Nordic nation with regard to the effects of European market integration. Denmark has a population density at a continental European level (see figure 8.1) while the other Nordic countries have small populations spread over large areas and a settlement structure based on rather few metropolitan nodes.

Although the other Nordic countries have small population bases, the areas are comparable in size to the rest of the EEA territory. The Nordic EFTA countries in 1994 had one-third of the EU plus EFTA land area and only five per cent of the population.

This geographical and demographic structure of the Nordic countries has important effects on how the integration process will affect internal regional balances. As mentioned, the population base is not strong enough to operate a multitude of metropolitan areas with a variety of service industries. Therefore, the capital region will tend to be the dominating growth pole of the countries. The reason is that the capital regions are the strongest metropolitan areas and much larger in size than the second urban region. Except for Sweden, the capital region is also the main gate for international connections, either by land, sea or air. Such a skewed internal regional balance is in contrast to egalitarian political programmes. The aim of regional policy is thus to try to prevent regional polarisation by stimulating growth possibilities of regional centres and non-central areas.

The differences in population density leads to different emphases in regional policy. In EU, the regions which have the highest priority in the transfers from the Structural Funds, the Objective 1 regions, have a rather high population density compared to the Nordic regions. Their problems are definitely not related to a lack of people. On the contrary, they may be regarded as having too many people relative to the production potential. For these Objective 1 regions, the problems are partly connected to the obsolete production structure, and partly to lagging welfare (low income, high unemployment). Regional policy is oriented towards these problems, the welfare indicators being the criteria for support, and restructuring being the aim of the support. Net outmigration is not a problem in itself. On the contrary, it may be highly recommended, as it will benefit more than just the migrants. As the problems of the regions may partly be due to a mismatch between the production potential and the size of the population, those left behind may also benefit from outmigration.

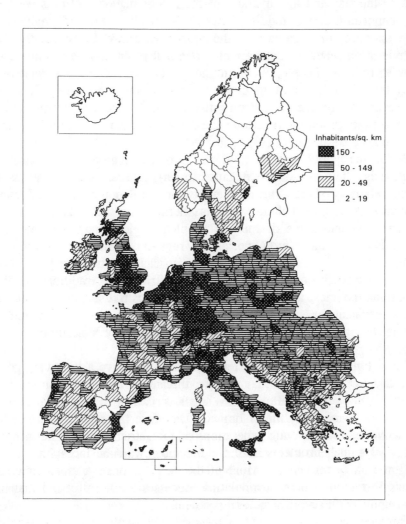

Figure 8.1. Population density, 1980
Source: Basprojektet 1990/91

In the low density Nordic countries, on the other hand, the net outmigration may threaten the basis of the peripheral settlements. A reduction of the population may weaken the social structure of the local communities, undermine the operating base for social and industrial services, reduce the size and the diversity of the labour market and thus weaken the opportunities for industrial activities. Such downward trends are now operating in large parts of the Nordic periphery. Consequently, net outmigration seems to be the major regional problem in a multitude of Nordic regions. People living in the peripheral regions are relatively well off when it comes to employment, income, and service access. However, partly due to a strong distributional policy, the regions still have serious problems regarding the future potential as operative communities.

While the population base in the problem regions of continental Europe is often strong, the distance to larger urban agglomerations is often relatively short. The problem is more often that the production structure is lagging behind, not peripherality as such. Thus, actions which may help some of the EU problem regions may not have the same effect in the Nordic peripheral regions. A number of projects directed towards modernising will have no chance when the population base is too weak. And as the Nordic countries have traditions of distributional policies, as well as using direct industrial support and public sector expansion as tools to reduce unemployment, the welfare indicators in the problem areas in the Nordic countries may look quite favourable.

Those living in the problem regions may often have a relatively good life, and yet the regions may still be threatened by depopulation. The small population base allows only small variations in the social arena. The scope of business life is limited. Therefore the region may not be attractive either for young generations or for entrepreneurs. The method for solving such problems must be directed toward business support as well as service provision. "More of the same" may be a more effective policy in regions where depopulating processes are the problem than in EU regions with severe structural problems.

This does not at all imply that the problems of the periphery are not related to structural issues. However it may give an explanation for the fact that regional policy in the Nordic countries is more directed towards firm oriented operating and investment support than most of the EU policy. In addition, the situation described here helps explain why the EU indicators of regional problem areas do not fit the depopulation problems of the Nordic support areas.

Since 1989, economic development has been somewhat turbulent in Europe, with weak growth rates in Western Europe and continued high unemployment rates. New market economies have emerged in Central Europe with varying perspectives for recovery and unemployment far above the Western European level, combined with a collapse in the Russian economy. As the Nordic countries have become more integrated into European economic development, they have also been affected by these trends. There are, however, important differences among the individual countries, both in trade patterns and in industrial structures, causing these countries to be exposed differently to impulses coming from the rest of the European economy (see Figure 8.2).

In Finland, a deep recession has taken place during the 1990s. In 1993 GDP was reduced by 12 per cent and employment by 18 per cent compared with 1990. The bottom of the business cycle was reached in 1994, and since then, a recovery has taken place, although at a rather moderate speed. As Finland traditionally has strong trade links with Russia, the collapse of the Russian economy caused Finland to suffer especially hard in the early 1990s.

Sweden also had recession problems in the early 1990s, although at a more moderate level than Finland. In 1993, GDP had been reduced by 5 per cent and employment by 12 per cent compared with 1990. The figures from the last half of 1994 indicate that the bottom was passed in winter 1994. In Sweden, much of the problems are linked to public deficits which call for a tough scaling down of public expenditures. This process is at present far from an end, and the budgetary cutback of 1995 may lead to continued recessionary effects in the following years.

Denmark has been through recession problems in much of the 1980s and is now on a more robust economic level less sensitive to business cycle problems. This does not mean that employment problems are non-existent. The unemployment rates in Denmark are on a relative high level (12 per cent in 1993 and 1994 according to national statistical definitions). But the GDP figures are showing a stable moderate growth, causing the employment to show only a weak yearly decline (the employment in 1993 is only 2 per cent below the 1990 level).

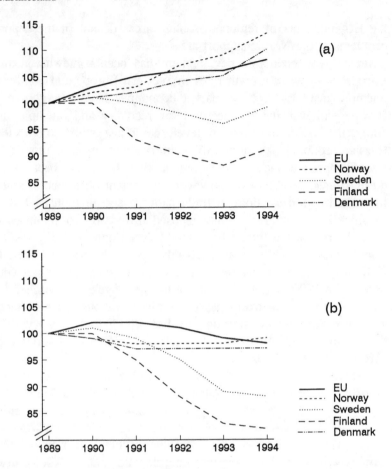

Figure 8.2. Real GDP (a) and employment (b) in the Nordic countries and EU, 1989–1994. Indexes, 1989 = 100
Source: OECD Economic Outlook. The 1994 figures are estimates made by OECD per November 1994

Among the Nordic countries, Norway has performed best during the first part of the 1990s, and also obtains the best level on the GDP and employment indicators. Of course, the oil income is a main factor behind this performance, making Norway one of the very few countries to fulfil the Maastricht criteria of acceptable budgetary policy.

The economic development in the last five years has created a somewhat new economic climate among the Nordic countries. The differences between the individual countries have been more visible, and the severe problems emerging in Finland and also in Sweden have led to a challenge to the traditionally distributional and welfare oriented policy. Regional policy also had to meet new economic challenges in the light of the economic problems, with the focus turning more toward local promotion even in the more centrally located regions.

8.4 Changing Industrial Structures

Technological development normally implies that the same amount of production can be done with less labour. Thus, the surplus labour force must be used for some type of productive expansion, if unemployment is to be prevented. Historically, this type of expansion has often been of a territorial nature, where more marginal agricultural areas were settled and long-distance emigration occurred as people sought new land.

In the last centuries, this extensive expansion has been replaced by an internal expansion, where the development of services and above all new manufacturing industries have created labour demand centres. These centres have been located partly in and near the older trade based towns, and partly in new resource based industrial sites (close to mineral ores, waterfalls, good harbours, etc.). In the Nordic countries, this development came somewhat later than in continental Europe, and it is not until this century that the manufacturing industry reached the same employment level as the primary industries (see figure 8.3). Later, the manufacturing industry also reached a level where the productivity growth caused reduced labour demand, and different types of services took over as the main labour demanding industry (including public services).

Employment in the primary sectors has been reduced gradually over the last decades. The manufacturing industry had growing employment up to the mid 1970s, then a stagnating phase set in, which after some years resulted in a reduction of employment. Services as a whole have been expanding, but this is a heterogeneous category where subsectors may have diverging growth trends.

This is more or less a common pattern of the industrial development in most countries. The countries differ in the timing of the phases and the level of employment in the industrial groups.

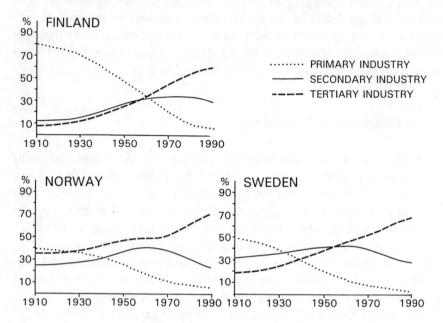

Figure 8.3. Employment shares of main industrial groups
Source: Basprojektet 1982; national statistics

As the industries have different location patterns, structural changes among industries impose important share-effects on the regional structure. The primary industries have for natural reasons a widespread regional pattern, and as they reduce the employment, the peripheral regions are strongly affected. For these reasons, actions which hamper the employment reduction in primary sectors, e.g. by support to small farmers and regulations to secure small boat fishing, are an important part of the total regional policy. For economic reasons, such policies would not be used on a level that would totally hamper the employment reduction within primary industries. Therefore, the perspective of the periphery is dependent of the regional ability to attract other industries to compensate for the reduction in primary employment.

Manufacturing had, from the beginning of this century, mainly an urban location pattern, due to the better transport and housing facilities around the old trade centres. However, important large industrial plants were set up at remote places, as cornerstones of new industrial sites. During the last decades, small scale industries were located in many rural regions, a development which was actively stimulated by the regional policy support schemes. At the beginning of the 1970s, the central regions still had a high share of manufacturing employment. Starting in the early 1970s, manufacturing employment in the central regions stagnated and declined. However, in more remote regions, manufacturing employment continued to grow. So, during the 1970s it seemed as if the regional policy was rather successful, and that the manufacturing industry could give the necessary employment stimulus to compensate for the reduction in primary industries.

After a while, however, the more remote manufacturing industry began to stagnate and reduce its employment as well. It seemed that it was, in fact, the central firms who made the quickest adaptation to the new technological and competitive climate, and that it was also the most labour intensive industries which were located in the remote regions (see Table 8.1). This implies that those manufacturing firms which are best equipped to meet the future challenges are centrally located, and those which have the greatest possibility for closing down and for rationalising are located in the remote regions.

Table 8.1. Newly created employment in manufacturing industry by production segment. Norway 1976–85, per cent

	The industrial concept based primarily on				
	raw materials	labour force	customer adaptation	high technology	total
Oslo region	3.4	41.6	46.8	8.2	100
Other large towns	5.9	59.5	28.1	6.5	100
Smaller towns	5.3	69.7	19.1	5.9	100
Rural regions	3.7	83.6	10.9	1.8	100

Source: Isaksen (1988)

The manufacturing industry has changed its position from being a centre of employment to being a source of future employment problems within the remote regions. Manufacturing of the future will increasingly demand good networking conditions and highly competent labour replacing routine labour. The future patterns of location will probably be more and more oriented towards urban agglomerations of a size larger than most Nordic centres.

As it seems difficult to rely upon the manufacturing industry as a future employment source, either on a national or on a regional level, the service sector will be the crucial factor for regional development. The manufacturing industry will of course be of importance as producer and income generator, but as the employment potential will be limited and the capital flows are more geographically mobile, the local spin-off from the productivity gain will be more dubious.

The service sector may be divided into two groups: those producing for firms and those producing for persons. The first group will normally have a strong centralised pattern of location, as it will generally be oriented towards technological and strategic business units. The latter will more or less reproduce the existing settlement pattern, with a slight tendency to be somewhat ahead in the centralising process.

Public financial policy will be of crucial importance for regional development. If a policy is geared towards stimulating personal purchasing power, consumer oriented services will expand, giving a more even regional development. If a policy is geared towards stimulating the economy of firms (so-called competitive power policy), then the centralising process will be speeded up. Since public activity constitutes a large share of consumer oriented services, the level and regional structure of public sector growth will be of major importance.

8.5 Changes in Public Policy

The Nordic countries have a reputation for a strong welfare policy with a more active public sector than in most other western countries. In reality, we find important differences among the Nordic countries. For some Nordic countries, the public part of the total economy is not so different from the average EU level. Figure 8.4 shows the public consumption and public expenditures, both measured as a per cent of GDP, in the Nordic countries and in EU-12, for the period 1970 to 1995 (both 1994 and 1995 are OECD estimates per autumn 1994).

Public consumption has a much higher level in Sweden and Denmark than the EU average. In Norway, the level of public consumption relative to GDP has been only slightly above the EU average up to 1980, and in Finland the level has been rather parallel to the EU average. The variance within the EU then implies that neither Finland nor Norway had a higher share of public consumption than that which was also common to the EU countries. Sweden and Denmark, however, had a significantly higher level of public consumption.

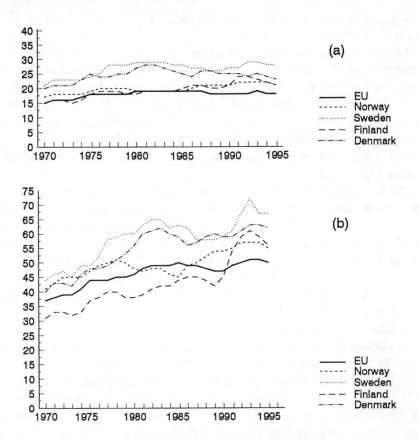

Figure 8.4. Public consumption (a) and public expenditures (b) in the Nordic countries and EU, in per cent of GDP
Source: Several editions of OECD Economic Outlook, OECD Main Economic Indicators and Statistical Yearbook Norway. Figures for 1994 and 1995 are from Economic Outlook, December 1994.

During the 1980s, the public consumption share stagnated in the EU countries, and then started a moderate decline. In all the Nordic countries the rising trend stopped in the beginning of the 1980s, and in Sweden and Denmark the share then declined significantly. After 1985, Denmark continued to reduce the public consumption rate, while Norway and Finland went into a new phase where public consumption expanded quicker than GDP. Sweden started to expand the public consumption share again from 1989 on. For the years 1990–1993, parts of the rising shares in Finland and Sweden are due to the falling GDP values, as shown in Figure 8.2.

Since 1993, all the Nordic countries have reduced public consumption shares, a trend estimated to continue also for 1995. For Sweden, the new state budget decided for 1995 indicate a more drastic reduction in public consumption than estimated in the OECD figures. At present, then, all the Nordic countries have public consumption shares significantly above the EU-12 average, but with declining trends. Thus, the gap between the Nordic countries and EU seems to be reduced at a rather high speed.

Public expenditures as a per cent of GDP evidence the same tendencies as public consumption. The shares were rising up to the first part of the 1980s. After a phase of stagnation they started to grow again in the Nordic countries while the EU-level continued to be stable. The levels were higher in Denmark and Sweden than in Norway and Finland. Finland has, in fact, had a lower level of public expenditure relative to GDP than EU up to 1990. The rise in the Finnish figures after 1990 is partly due to the falling GDP level and partly to the rising unemployment payments.

Traditionally, the difference between EU and the Nordic countries regards consumption rather than transfers. It is Denmark and Sweden, not Norway and Finland, that have higher consumption rates than the EU-average. The main difference between the Nordic countries and EU is the public response to social problems, since both public consumption and transfers are used to counteract problems in the market economy. Budgetary constraints in the Nordic countries are now hampering this type of response to the problem.

The post-war period up to the beginning of the 1970s was characterised by stable economic growth. The growth could be used to build welfare, and the public sector could take responsibility to both secure a distributional policy and to secure growth in average welfare, by continuously expanding both the quantity and the quality of public services. When the economic growth rates were weakened in the beginning of the 1970s, this was looked upon as something that the state could and should counteract. The growth rate of the public sector of the Nordic countries then continued at about the same speed as before, while the EU countries made a much quicker response to the new economic climate. At the end of the decade, it was realised that the stagnation tendencies of the manufacturing industry were structural rather than cyclical and therefore should not be counteracted in a traditional Keynesian way. Actions were taken to limit the public growth, and to tighten the economic policy. In the 1980s, there were again closer connections between private and public growth rates. No wonder then that in the 1970s, public employment took the lion's share of the total employment growth, and that this decade became special.

The public growth was organised according to sectoral targets, where some predefined service levels could be reached within a certain time perspective. These service targets set maximum travelling times for each household to reach a school, or a health institution, and established minimum quality levels in each institution. Thus, those regions which originally were most below such targets, attained the highest growth rates of services. Thus, the public sector growth was much stronger in the remote regions than in the central ones.

This unusually strong relative growth of public employment in the remote regions, caused a stabilising mobility effect on the settlement patterns in the 1970s. As public growth slowed down again in the 1980s, the centralising mobility patterns from the 1960s reappeared.

In the 1980s, international competition grew stronger, and then the attention of the government was geared towards strengthening the competitive industrial power of the nation. This dictated that distributional policies received lower priority on the political agenda. This resulted in a tighter economic policy, which reduced the growth both of public services and personal purchasing power. More efforts were used to get better productivity in the public services. And as the regional targets were becoming less important compared to the overall competitive power, the result was a more centre-oriented growth of public sector. Both the lower growth rate and a changed regional profile resulted in a serious weakening of the counter-centralising effects of public activity.

The new economic problems from the end of the 1980s involved both central and remote regions, with growing unemployment also in larger cities. The new expansion of public expenditures did not have the same periphery oriented profile as in the 1970s. In addition, the new labour market problems also tended to undermine the traditional legitimacy of periphery support, as the problems were now visible in the central regions as well.

8.6 Perspectives and Conclusions

Except for a temporary break during the 1970s, the net migration movement in the post-war years has followed a centralising pattern. The migration pattern of the 1980s has strong similarities to the pattern of the 1960s.

The centralising movements in the 1960s were not regarded as threatening for the settlement pattern. Even with high net outmigration, the natural population growth of the remote regions was strong enough to uphold and often even expand the population numbers.

The fertility rates were, however, reduced during the 1970s. When the centralising net migration pattern was restored in the 1980s, the natural population growth was no longer large enough to prevent population decline in the remote regions. Thus, the effects on the settlement pattern became more serious. In the 1960s, nearly every municipality had population growth. Most of the peripheral regions now face population decline. The exceptions are mainly the larger college centres.

The two main factors behind the stable development of the settlement pattern during the 1970s, the periphery dominance of manufacturing and public sector growth, have now vanished. The result is a situation in which the periphery regions must face the problems of declining employment in the traditional industries, without counteracting sectors filling the gap. The perspectives therefore imply a continuing centralising trend.

Both from an industrial and demographic viewpoint, there are reasons to predict that the centralising process will be strengthened in the future. Firms and households consider both the actual condition and the perspectives of a region before settling. Network conditions, service levels, etc. will be important factors. A self-accelerating processes may develop in which the losing regions will lose more and the strong regions will attract even more firms and households. The driving forces behind this development are partly structural, and it seems unlikely that the process will slow down without conscious actions.

Some of the remote regions of today may turn out to have production factors which will make them attractive regions. Such factors can be on the environmental side, for example, low-chemical products from small scale farming is expected to have a higher demand. Fishing resources may also have a strong demand in the future.

We also see a growing interest in regions which are adjacent to, but still outside the central towns. In addition to the main trend of centralisation, we have a decentralising trend within short travel distances from the larger towns. Thus, the centralisation process is a process between larger regions, where the tendency may be different within such larger regions.

The conclusion that stronger centralisation seems to be the result of strong interregional competition and reduced geographical shelters, may be converted also to the European level using a rather macro-geographical level. The European integration process would therefore give the same effect for all remote regions without strong metropoles. The appeal of making regions which are moderately distant from metropolitan regions may be the dominating trend reflecting counter-urbanisation forces on the micro level.

In fact, although formal barriers are being removed in the internal European market, a lot of informal barriers remain. Different language, traditions, and environmental factors keep international mobility on a much lower level than internal mobility. Through time, however, such barriers will tend to weaken, as well.

The experience from the 1970s demonstrates that public policy, if strong enough, can affect migration streams. However, since the underlying redistributive forces seem to be rather strong, regional policy will most often be too weak to alter the direction of the process.

Maybe the best approach to policy is to accept that a region needs to have high level centre facilities within reach in order to cope with the general integration processes. For large parts of the Nordic territory, this will be difficult to obtain. An alternative strategy must then be based on the premise of a remote location. This means that such regions should not join the race of integration, where competition from the larger centres will be growing. Rather, the special status of the peripheries in the high North, combined with access to natural resources, clean water and good air quality (as long as we avoid severe natural damages from industries and nuclear accidents) should be regarded as assets for regional survival.

The point here is not to design specific types of policies. The point is that such policies should take the unique features of the different regions into account. A bureaucratic standardising of regional support schemes on the EU or EEA level should be loosened up to avoid greater problems in both regional policy and regional development. If regional sheltering is difficult to operate within the framework of the EU rules, then the EU integration will easily create more costs than gains for large parts of the Nordic territory.

References

Basprojektet, *Regional utveckling i Norden*, Bi-annual reports, Nordic Council of Ministers.
Isaksen, A. (1988), *Næringsutvikling i sentrum og utkant*, ØF-rapport nr. 5:88, Østlandsforskning, Lillehammer.
Nordic Council of Ministers (1991), *Nordic regions and transfrontier cooperation*. Copenhagen.
OECD, *Main Economic Indicators*, Monthly editions, Paris.
OECD, *OECD Economic Outlook*, Semi-annual reports, Paris.

CHAPTER 9

The Fall and Revival of the Swedish Welfare Model: Spatial Implications

Stephen F. Fournier
Kungl Tekniska Högskolan
Stockholm

Lars Olof Persson
Kungl Tekniska Högskolan
Stockholm

> "Everyone should have the right to childcare, schooling, healthcare and eldercare. That is the way we want it in Sweden. The turnout in this year's election clearly shows that we are willing to fight for the protection of the Swedish model. The general welfare system should be developed, not diminished." (Faxservice of *Dagens Nyheter*, October 9, 1994.)
>
> Ingvar Carlsson
> Prime Minister of Sweden
> October 9, 1994

9.1 Introduction

The most rapid expansion of the welfare state in Sweden took place between 1960 and 1985. In the mid 1980s, public services provided about one third of all jobs in most labor markets; in some regions as much as 45 per cent. Standardized formal education in Sweden is aimed specifically at providing well-educated persons for the public sector which results in very little labor mobility between the public and private arenas. More than two thirds of those with higher education work in the public sector. Consequently, in most regions, there has been little room for the growth of private personal services and the private sector is, as a whole, considered to be under-supplied with highly-educated people. In addition, the relatively flat wage structure embedded in the Swedish welfare model provides little incentive for movement between public and private sectors. This is one way that the welfare state, which explicitly and actively promotes regional equality, has also generated certain regional problems. Its role in this sense has largely been implicit, indirect and not widely recognized.

The Swedish welfare model is currently facing restructuring and deregulation causing changes in virtually every labor market. Cut-backs in public services are leaving more room for private entrepreneurs. Internationalization is leading to increasing wage differentiation. Low paid service jobs are emerging in response to lower personal taxes and the withdrawal of some public services. The regional consequences of these changes are far-reaching. Equality between regions, an important element implicit in the Swedish model, is being altered. Decreased decentralized demand for highly-educated public-service labor coupled with increased wages for qualified labor in private industry is moving labor to larger urban labor markets (cf. Axelsson et al., 1994). Although the industrial sector carried increasing costs to finance the public sector which may have hampered growth, it has also been recognized that the business sector benefited from the stable institutional framework associated with the large public sector. Correspondingly, the present period of restructuring means increasing uncertainty for industry with a number of negative impacts such as low investment rates (SOU 1992:19).

In this chapter we present an historical overview of the Swedish welfare system and describe some of the recent changes, especially as they impact upon regional labor markets. We discuss and analyze these developments with focus on the role of the public sector and the welfare state as indirect regional policy instruments. Our purpose is to offer a review as a starting point for further research as there is much work to be done by regional scientists in analyzing the large and multifaceted role played by the public sector and the welfare state in a number of regional processes.

9.2 Evolution of the Welfare State in Sweden

9.2.1 A Long Period of Growth

The evolution of the Swedish welfare state began in the 1930s under leadership of the Social Democratic party. An informal alliance developed among organized labor, the state and major private capital. One over-riding goal was the equalization of living conditions in different social strata, popularly expressed by the metaphor of "The People's Home". The program was based on large scale intervention by central government in the economy, centralism and uniformity.[1] High levels of international competitiveness within industry provided expanding economic resources for the nation which created financial possibilities for an increased role for the public sector and for an expanded role for public responsibility. Numerous public sector institutions were created for the organization and production of public services. The legitimacy and widespread acceptance of the model is clear from the unbroken rule of the Social Democrats from the 1930s through the 1970s (Myhrman 1994; Persson & Wiberg 1995).

The real take-off for the welfare state as a producer of services came in the 1960s when it was underpinned by strong economic growth.[2] Welfare benefits accrued not only to the poor but also to the middle class. During the 1970s the welfare state created reasonably paid jobs for a large share of women. In practice, the traditionally informal and unpaid female jobs within families were transformed into formal jobs in the social and health care sectors. Private entrepreneurship and competition was largely excluded not only from education, health care, social care and associated sectors, but also from urban and regional transportation, telecommunications, domestic airlines and railroads. Private initiatives in many sectors (such as housing) were restricted and regulated. As a general rule, basic public services were offered free of charge or strongly subsidized. To some extent, Sweden became a model state as a high private living standard (in spite of high personal taxes) was accompanied by probably the most ambitious (and expensive) public safety net in the world (Ginsburg 1992).

The labor market and economic policy programs of the 1960s focused on stimulation of further economic growth by supporting the transfer of people from low productivity sectors such as agriculture and forestry to dynamic manufacturing industries. This structural change had spatial consequences. Explicit regional policy formulated in the mid 1960s concentrated on incentives to industrial locations in less developed regions (primarily in the northwest) which were experiencing a net loss of labor. The historically successful export orientation of large manufacturing corporate firms was a necessary precondition for steady growth.

---

[1]This was later to be termed "The Strong Society", a positive concept.

[2]This period became known as "The Record Years".

This small group of large Swedish-owned companies in the context of a relatively small national economy contributed to an impression of a harmonious alliance between the state, private industry and the labor force. However, during the 1970s the Swedish economy faced problems of stagflation, unemployment and high militancy from labor unions which put pressure on the government to institute a Keynesian defense of its welfare goals. Maintaining a low open unemployment rate became a key strategy. The iron and steel industry and shipyards encountered strong international competition and faced decreasing market shares. It became evident that the cost of production in Sweden had risen above the level of international competitiveness. This was due in part to the resource demands by the public sector but also because labor unions had negotiated successfully for increased wages based on a principle of solidarity, which was interpreted as the same wage for the same job in each industrial branch regardless of location or the firms' ability to pay. Although this impacted strongly on less productive units in peripheral regions, regional policy had simultaneously moved to a distributional function and a variety of means were available to compensate for plant closures or other dramatic reductions to labor. It can be argued that regional equalization policy became superior to economic stimulation for further economic growth.

Swedish regional policy since the middle of the 1960s can be divided into four phases (SOU 1989:65):

- From 1965 to 1972 policy was focused on modernization of lagging rural and urban areas located in the sparsely populated northwest area. The key concept was "balanced regional development". Efforts were focused on support to manufacturing, industrial firms and basic services along with investments in transportation facilities and improvement in the public transport systems.
- The period between 1972 and 1976 was characterized by municipal reform which was accompanied by a rationalistic plan for the further development of the centers of the 284 municipalities and relocation of administrators from a number of governmental authorities out of Stockholm to medium-sized cities.
- The phase between 1976 and 1985 evidenced a shift from the principle of a ruling central government to stress on the mobilization of local resources. During this period the plan for the future central place structure was discontinued.

- After 1985 there was a focus on improvement in infrastructure for rapid transportation and communication as well as the further development of professional and technical competence and cultural institutions in lagging regions. Major new measures included the decentralization of higher education and special education and technology diffusion programs. Support was also provided to service firms which relocated to low-employment regions.

Subsequent regional policy has been developed to harmonize with EU policies for regional development. In 1994 the new regional policy is expected to promote "viable regions" along with "equally good living conditions", with each region contributing to the growth of the national economy (Government Bill: Regeringens proposition 1993/94:40). The peripheral northwestern region, dominated by sparsely populated areas, however, is still given highest priority.

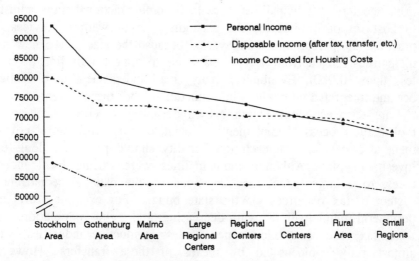

Figure 9.1.   Income and Buying Power
Source: NUTEK 1994

Income data seem to verify that this direct regional policy has been very successful (Figure 9.1). However, there are other explanations for the spatially equal distribution of buying power. Figure 9.1 shows that although personal incomes differ widely across regions in Sweden, the impacts of taxes, transfers and housing costs make the resulting distribution much more equal. It is becoming more accepted that the impact of direct regional policy (explicit aid to economic development in specified regions) on equalizing living conditions is of much less importance than indirect regional policy, i.e. the different programs within the public sector as a whole. In the government's bill on regional policy in 1994, it is stated that the spatial impact of different subsectors of the public sector should be considered and respected in public decision-making.

9.2.2 The Local Arena

To enable local areas to fulfill uniformity principles in service supply and welfare provision, reforms were introduced in the early 1970s that merged a number of municipalities. This reshaping was based on principles of central place theory with the explicit recommendation that each municipality should form a functional unit – in terms of labor, housing and service markets – embracing both less-developed rural regions and urban centers. Municipalities were to become functional units adjusted to host responsibilities for the expanding set of welfare services. In practice, the preconditions still varied because the size of the new municipalities ranged from 3,000 inhabitants to almost 1 million (average less than 20,000). Eventually, many small independent municipalities became integrated with others into functional labor market areas.

Long-term economic planning by the central government became a tool in the process of implementing social reforms. The central government also decided that each municipality should prepare a long-term investment plan. Although municipalities were raising local income taxes, to a large extent they depended on the national redistribution system of tax resources via the state budget. For example, rural and peripheral municipalities with decreasing and aging populations and weak tax bases (a predominance of less productive industries) were almost fully compensated by means of these transfers. However, detailed rules restricted how municipalities' financial resources were used concerning the standard of schools, social welfare, public housing etc. In retrospect, at least, the general impression is that local governments all over the country, even those dominated by conservatives and liberals opposing far-reaching state intervention, developed into satellites of the central government. The continuous growth of resources transferred to municipalities was one precondition for this.

Our conclusion is that the growth of local public services became the strongest instrument for a spatially equal distribution of employment and welfare in the period between 1960 and (at least) 1990. Compared to specific regional policy (i.e. incentives to industrial firms located in, or relocating to the northwestern aid area), the spatial impact of the distributive and compensatory function of the multidimensional public sector has been much stronger.

9.2.3 The Changing Public Sector

The Swedish welfare model evolved systematically over four decades following the end of World War II with broad approval among voters. The Social Democrats were largely unthreatened in government for an unbroken 44-year period ending with the formation of liberal-center-conservative governments in the late 1970s, and even then the legitimacy of the model was still not strongly questioned. It was only after 1990 in a period of slow productivity growth, increasing state budget deficits and a deep recession that the new broad liberal-center-Christian democratic-conservative government expressed and revealed preparedness to revise some of the ideological base of the welfare model (Eklund 1994; Berglund & Persson 1994).

As we have shown, Sweden now stands out as a country with a comparatively even distribution of disposable income. This is due both to a relatively even distribution of gross incomes and a redistribution via the tax and transfer system. In the mid 1980s, the welfare state redistributed 53 per cent of the differences in incomes, higher than any other country in Europe. Income transfers to households amounted to one fourth of GDP. Consequently, the poverty rate after redistribution was less than in most other European countries. This has, however, put a heavy burden on tax payers and in the late 1980s there was rather broad consensus in Parliament that the tax rate had to be lowered in order to avoid further distortions in the economy.[3]

---

[3] However, it is not only the size but the efficiency of the transfer system which is in question. Söderström (1988) estimates the same equalization could be achieved with transfers worth only 3 per cent of GDP. Hence, in theory, substantial cut-backs need not have negative effects on income distribution although in practice it is widely questioned whether this is possible.

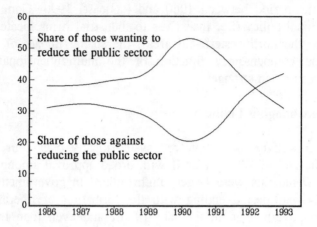

Figure 9.2. Opinion on the Size of the Public Sector: 1986–1993
Source: Holmberg 1993

At that time, support had decreased both for further expansion of the public sector and its concomitant increase in taxes. Changes in the international arena, with the fall of neighboring socialist states in Eastern Europe, contributed to an increased resistance to central planning and state intervention. As shown in Figure 9.2, the proportion of people demanding reduction in the size of the public sector increased from about 38 per cent in 1986 to a peak of 55 per cent in 1990. The election to Parliament in 1991 reflected these shifting attitudes. Around 1990–1991, less than one fifth of the voters objected to a reduction in the public sector. The recession which started at almost the same time, however, had an important impact on public opinion. Three years later support for the public sector was as strong as it had been in the mid 1980s. This was certainly expressed in the election in September 1994, which brought the Social Democrats back to power.

9.2.4 Swedish Versus Other Nordic Welfare Growth

The development of the welfare state in Sweden is closely linked both to the four decades of Social Democratic power in parliament and the relatively stable economic growth in the post-war period (relative to most other European countries). Ginsburg (1992) argues that in most countries party politics has a very strong influence over the evolution of the welfare state. Thus, comparative political differences account for most of the differences in welfare expenditures between countries. Alber (1983) compared social security expenditures and the political structure of 13 Western states between 1949 and 1977. Not surprisingly he found that 'left' cabinets tended to increase the social expenditure ratio more than other regimes. The restraint in welfare state growth in the 1980s has been more marked in countries with conservative governments. However, it must be stressed that the development of the welfare state must always be explained with respect to the social and political history of each state.

Esping–Andersen (1990) used seven indices for measuring the accessibility, coverage and redistributive impact of benefit systems along with four indices of a "welfare regime". Applying these indices to OECD states he found close correlation between the nature of the benefit system and the type of political regime. In this study, special emphasis is given to the concept of de-commodification, i.e. the extent to which social security benefits are linked to labor force participation. Esping-Andersen describes three clusters of OECD welfare states according to their rank order using this measure. Sweden, Norway, Denmark and The Netherlands exhibit the highest ranks while Anglo-Saxon states such as Australia, the US and the UK show the lowest. Finland, France and Germany are located between these groups.

By aggregating a number of political indices, Esping-Andersen identified three types of political regimes with differing types of welfare systems; strong liberal (e.g., Australia), strong conservative (e.g., France) and strong socialist (all Nordic countries and The Netherlands). The liberal regimes are characterized by a relative absence of working class mobilization. Conservative welfare states emphasize social insurance over both means-tested and private benefits, but in forms which tend to maintain class distinctions. Socialist or social democratic states have achieved more universal and the most class-redistributive benefit systems, modifying class structure but not removing it.

These results indicate that the link between decommodification and a socialist regime is strong in Sweden, Denmark and Norway but clearly works less well for Finland. This reflects the fact that the Nordic model is not entirely the same in each country. Finland has followed a path similar to Sweden but started much later. Moreover, regional policy in Finland (and Norway) has been more directed to rural residents, especially farmers. The spatial implications of these welfare states are also different. In Sweden and Denmark, significantly more than 30 per cent of employment is within public employment showing the dominance of the use of the public sector in balancing regional development in these countries. In Norway the corresponding figure is 27 per cent and in Finland only 22 per cent (Yearbook of Nordic Statistics 1994).

9.2.5 Revenues and Redistribution

The core of revenues to Sweden's state budget comes from taxes on labor and consumption and a progressive income tax is one element of the redistribution system (see Table 9.1). Due in part to this, there is wide variation in how much the population and enterprises in each municipality contribute to state revenues per capita. In 1992, Stockholm contributed most; about 32 per cent above the national average. Municipalities contributing least were, to a large extent, located in the northern periphery. Their contribution is almost 30 per cent less than the national average. The reason for the high number for Stockholm (and to some extent Gothenburg) is the relatively high wage level and the high rates of labor market participation, which generate both high wage taxes and value added tax.

Table 9.1. Governmental revenues, and expenditures, Sweden 1991/92 (per cent)

Tax Source	per cent	Expenditure	per cent
Wage Tax	51	Transfers to Households	47
VAT	25	Government Administration	18
Individual Income Tax	5	Grants to Municipalities and	
Other	19	County Councils for Local Services	17
		Labor Market Policy	10
		Interest Subsidies in Housing	5
		Support to Agriculture and Other Industries	3
Total	100	Total	100

Source: NUTEK 1994

With Swedish taxes amounting to about 55 per cent of GDP, there is widespread opinion that the total tax pressure cannot be increased without introducing severe distortions to the economy. In most other countries, taxes amount to less than 40 per cent of GDP and in the US only 30 per cent. In 1992, the corresponding figures in Denmark, Norway and Finland are 50, 47 and 47 per cent respectively (OECD 1994b).

There is also strong variation among municipalities in terms of total governmental expenditures (cf. Table 9.1). A number of municipalities dominated by military expenditures or with concentrations of governmental administration receive more than 35 per cent more while certain suburban municipalities in metropolitan regions receive less than two thirds of the national average. This is due largely to the lack of governmental work-places in these municipalities; differences in welfare services and payments do not account for the extremes of the spatial variation in total expenditures. Still, the general picture shows that lagging regions are favored by indirect regional policy.

9.2.6 Cut-backs

Cut-backs have recently been substantial in sickness payments, grants to health care, pensions and subsidies to municipalities for local services and defense (OECD 1994a). Despite a net reduction in the state budget in recent years, some programs continue to expand. The largest of these affect the labor market due entirely to the present high levels of unemployment. Programs with increasing costs in 1992/93 include primarily labor market policy, but also grants to higher education, interest subsidies to housing, reception of refugees, research and education, and transfers to families with children and to the handicapped.

This tendency over the past few years to cut public spending and reduce state intervention is certainly not unique to Sweden. Castells (1989) argues that the 1980s witnessed a general change in the pattern of state intervention, with emphasis shifting from political legitimization and social redistribution to political domination and capital accumulation. This is evident in several mechanisms that express the new form of state support, including:

- deregulation of many activities, including relaxation of social and environmental controls in the work process;
- shrinkage and privatization of productive activities in the public sector;
- regressive tax reform favoring corporations and upper-income groups;

- state support for high-technology R&D and leading industrial sectors which form the basis of the new informational economy. This support usually takes the dual form of financing infrastructure and research and favorable fiscal policies;
- shrinkage of the welfare state, with variations within and between countries according to the relative power of affected groups;
- fiscal austerity, with the goal of a balanced budget, and tight monetary policy. (op cit, 25)

It is clear that under the broad right-wing government in Sweden in 1991–1994, almost all of these issues were given priority (OECD 1994a):

For *transfer programs* the major impetus was to lower compensation levels. Interest subsidies were cut within housing policy. Sickness benefits were cut in several ways: compensation levels were reduced from 100 per cent to a low of 65 per cent and a high of 90 per cent of regular salary. A one day wait before benefits are paid was introduced. Responsibility for payment of these benefits was shifted from the state to employers. Altogether this has reduced absenteeism. This was certainly also reinforced by the current labor market situation.

To cut increasing costs for *labor market policy,* unemployment insurance now covers only 80 per cent of regular wages. Relief work wages have been reduced by 10 per cent. The pension system has also been changed, among other ways, by reducing the possibility for early retirement.

Most social services are produced by municipalities and counties. As mentioned, central government had strong influence on the allocation of resources in each municipality. The new trend is toward stronger independence, reinforced by the way resources are now transferred to local authorities, i.e. largely in the form of lump-sum payments. This leaves more room for *local priorities*.

Local priorities, within certain limits, also impact the local income tax rate. There is a tendency for high-income municipalities to keep local taxes low, while low-income municipalities usually have to finance expensive social programs by increasing the tax level. The rapid shift from an overheated labor and housing market to a reverse situation caused a number of municipalities to experience financial crises.

The reorganization of social services in the first years of the 1990s has followed three steps; first there has been a move to try to improve efficiency within the existing organization, secondly service producers have been allowed or forced to compete, thirdly privatization has been introduced. Private entrepreneurs are now allowed to compete in the fields of technical services, child-care, homes for the elderly and health services. Private schools are operating with 85 per cent of their costs covered by the government: as long as they follow the core of the national curriculum they are free to choose their own profile. In 1994 it is estimated that 20 per cent of the services financed by local governments are produced by private firms. Only two years ago the corresponding figure was a scant few per cent. The principle of free access to all social services is being abandoned and fees have been introduced for many services (DS 1994:16). Altogether, we anticipate seeing increasing differences among municipalities due to local policies regarding service subsidies and accessibility. The general principle still remains, however, that there should be public control of the quality of services and a regulation of fees. For example, the present government has declared a preparedness to intervene if tendencies toward segregation emerge in the schools.

9.3   Recent Social Developments

Reduced income taxes and public services, privatization of public services and government owned companies were some of the elements at the core of the economic policy program launched by the government immediately after election to Parliament in 1991. It is almost ironic that the program was entitled "A New Start for Sweden" as within a short time unemployment increased to levels not experienced in Sweden since the 1930s. Swedish export industry encountered severe problems due to high costs and the budget deficit increased rapidly. The offensive neoliberal policy program had to be moderated. Resistance to cut-backs in the welfare state increased because they contributed to unemployment and social inequalities.[4]

9.3.1 Employment

An economic upswing in the Swedish economy in the 1980s increased demand for labor in all regions in both the private and the public sectors. Total employment increased each year during the 1980s as did the female participation rate – much more than in most other industrialized countries. Among families with children it became almost a rule that both parents held jobs in the market.

This boom did not increase jobs for immigrants, however. Approximately one-fifth of Sweden's labor force is now comprised of first or second generation immigrants and during this period many moved away from the labor market. This is explained in part by the fact that immigrants are over-represented in those professions where early retirement is more common. The proportion of immigrant families with both spouses working decreased. Thus, for immigrants, employment problems actually started during the economic boom of the 1980s.

In the early 1990s demand for labor decreased in most sectors and unemployment increased with the highest rates among young people, immigrants and single parents. In households with two adults it is once again becoming less common that both are economically active. Immigrants are less likely to enter the regular labor market and also less likely to have access to labor market policy measures. The proportion of economically active handicapped people is also decreasing. The number of unemployed without unemployment insurance is increasing.

In addition to open unemployment, there are increasing numbers of persons with weak links to the labor market – underemployed and those in labor market programs. These groups outnumber by far the number of unemployed.

---

[4]Parts of this section summarize information from a recent survey of changing living conditions between the 1980s and the 1990s (Social Rapport 1994:10). The survey focuses on problems affecting children, single parents and immigrants. The spatial dimensions of these changes are not reported but it is implicit that most "new" problems have appeared in urbanized regions. Although we present some analysis of this report, much more needs to be done to understand fully the regional implications of these social changes.

9.3.2 Economic Resources and Poverty

Although personal incomes increased in the 1980s, so too did differences between socioeconomic groups. Across all socioeconomic groups and in private as well as public sectors, men had larger income growth than women and wage differences between men and women with the same level of education and work experience increased. From the mid 1970s to 1991, the proportion of families below the poverty level was reduced by half.

In the early 1990s, as many as 500,000 children (more than 25 per cent) were living in families with reduced economic circumstances. Age, gender and class also influence income, and low incomes are numerous among the young and the very old with many living close to the poverty level. Disposable incomes below the poverty level are becoming more common among blue collar workers. Although more men than women are below the poverty level, larger numbers of women are closer to the border of being classified as poor than men. Poverty is more common among foreign citizens.

Income transfers reduce the impact of poverty, first of all among the elderly but to some extent also among the young. The share of poor single parents was reduced in the 1980s. As much as 80 per cent of single parent families are moved out of poverty by social transfers (i.e. regular programs, not social benefits). Although the proportion of clients receiving social benefits remained stable throughout the whole postwar period, this changed during the economic boom of the 1980s as the number of households receiving benefits actually increased. Costs increased almost entirely due to increased payments to immigrant households. This was a reflection of both increased refugee immigration and the formal restriction that refugees waiting for asylum are not given working permits.

Among those qualifying for social benefits a majority has weak attachment to the labor market. However, only a minority rely on social benefits for long periods; most often families need temporary economic support. During the 1980s, the proportion of elderly people receiving social benefits was reduced while the proportion of immigrants increased. Until the mid 1970s, elderly people constituted more than 15 per cent of those receiving social benefits. In 1994 the corresponding figure was 7 per cent.

9.3.3 Housing

According to most indicators, housing standards in Sweden are good and most people live in modern and commodious flats almost all with housing contracts. Thus, most 'classic' problems associated with housing have been eliminated. In the 1980s, overcrowding continued to decrease and class-related differences were reduced in terms of housing quality. In the late part of the decade, however, young people faced difficulties entering the housing market, especially in metropolitan regions.

Housing segregation, both socioeconomic and ethnic, increased during the 1980s, mainly in metropolitan regions. The number of 'mixed' housing districts decreased after 1985. An increasing number of communities in the periphery of metropolitan regions have overrepresentation of immigrants and groups with meager general resources. Many of these areas are deficient in their built environment and provide poor services and maintenance. People in these areas report dislike and even fear about their environment. For some families, the high unemployment and increased economic pressures of the 1990s has contributed to reducing housing standards.

Deregulation of the housing market, reduced subsidies and market-orientation of rents is already influencing families as they attempt to maintain their housing standards. Young people, families with children and single parents are especially sensitive to changes in incomes and costs. On average 20 per cent of disposable income is used for housing costs in Sweden, more than in most other countries in Europe. There is evidence that recent immigrants and young people now face increasing difficulties entering the housing market. The homeless are also becoming more visible in the larger cities with one estimate that they now exceed 10,000 in the country as a whole (Socialstyrelsen 1993). This is slightly more than 0.1 per cent of the total population.

Explicit housing grants are provided mainly to families with children. For single parents, on the average, 40 per cent of housing costs are covered by such grants. Since demand for such grants is increasing while public resources remain the same, it is likely that many families currently receiving grants will face reductions within the years to come.

In summary, housing segregation is expected to increase as a consequence of high unemployment, reduced economic resources, refugee immigration and reduced subsidies within the welfare model. It is likely that differences between attractive and less attractive housing areas will increase further in metropolitan regions. In Sweden, the classic problems concerning housing, overcrowding and low standards, have been changed and replaced by problems concerning the housing environment.

9.3.4 Results: A New Set of Spatially Diverse Problems

This overview of recent social developments clearly shows the emergence of a new set of economic and social problems whose roots can be traced to the rapid changes of the 1980s. Two different sets of explanations have been offered. One stems from the weak financial policy, deregulation of the capital market and tax reforms. The other focuses on the continuous growth of the welfare sector which acted to hide the emerging problems for some time (Edin & Holmlund 1994). Previous regional problems were more readily identifiable as for example the difficulties with the sparsely populated and economically depressed northwest area. These problems lent themselves to longer-term directly focused regional policies. The new problems are much more diverse both spatially and across sectors. Intraregional problems are appearing in metropolitan areas. The seriousness and rapidity of their development has moved regional policies into a reactionary mode which does not go to the root of the problem. The challenge will be to balance both short and long-term policies into an affordable and effective force.

9.4 Government Programs with Spatial Implications

9.4.1 Typologies

We contend that, in a country such as Sweden, total government spending is of major importance for understanding regional development and spatial processes. However, programs aimed at various sectors evidence differences both in spatial impact and also in terms of the intention to contribute to regional policy goals. Accordingly, we make distinctions between four types of policies (Table 9.2).
- *Primary* regional policy has the explicit and primary goal of promoting regional development in selected regions. In terms of monetary

resources it is a small fraction (less than one per cent) of total governmental resources with a spatial distribution.
- In *secondary regional policy*, the spatial dimension is important, but it is given lower priority than other goals.
- The third group of programs have *distinguished spatial implications* but are not guided explicitly by regional policy goals. This means that some implications can be in synergy, some in conflict with regional policy goals.
- *Other* policy programs are not expected to vary systematically between regions.

Each of the more than 600 items in the state budget can be classified according to how they fit into these four types. Additionally, each item can be classified in terms of its function in a regional context. We find that many of these programs can be readily clustered: some are best explained as reactive to current problems, others intend to enhance the technological level and the competence level of the labor force, others aim at providing basic services in the regions, etc. In Table 9.2, five such program clusters are classified in terms of their temporal orientation, i.e. whether they are primarily oriented towards current labor market problems or oriented towards anticipated future problems.[5]

- *Response to Social Problems* covers 16 per cent of total expenditures. All programs included in this cluster are intended to compensate for acute problems in local labor markets. Approximately 60 per cent of all such compensatory measures are paid directly to individuals, 30 per cent to municipalities and the rest to business firms. Primary regional policy is included in this cluster. Today, labor market policy is the largest program demanding resources which are almost 20 times bigger than the primary regional policy. There is, not surprisingly, a certain correlation between the spatial outcome of this entire cluster and the primary regional policy.

---

[5]This definition of clusters is made for analytical reasons and does not reflect the sectoral division of responsibility between ministries. From the point of view of regional policy, it could be argued that the efficient regional coordination of sector programs is counteracted by the traditional division of labor between ministries.

Table 9.2. Policy goals and clusters of government programs in Sweden

Policy Type	Policy Focus	Total Expenditures (%)	Timeframe of Expected Impact
Direct policies			
Primary Regional Policy	Response to Social Problems	16	Immediate
Indirect policies			
Secondary regional policy	Research, Education & Culture	11	Long range
Implicit regional policy	Administration, Health and Infrastructure	19	Medium range
	Housing, Energy and Defense	11	Medium range
Other policies	Transfer Programs	43	-

Source: After SOU 1989: 65

- *Research, Education and Culture* accounts for 11 per cent of the total state budget. Education alone takes two thirds of this sum. The spatial concentration is obvious since almost all the resources for higher education, research and culture are directed to a limited number of local labor market areas.
- Programs within *Administration, Health and Infrastructure* account for 19 per cent of total expenditures. The major programs are administration, health and social care, infrastructure maintenance and support to agriculture. This cluster contributes substantially to job generation in the public sector. The capital region of Stockholm stands out as the most important recipient of these programs because of the concentration of central administration. In infrastructure maintenance, peripheries in northern Sweden receive three or four times as much in resources per capita as more densely populated areas.
- *Housing, Energy and Defense* accounts for 11 per cent of total expenditures and consists of the defense sector and the subsidy system

within housing policy. The spatial pattern is very biased towards a few municipalities with important functions within Swedish defense.
- *Transfer Programs* account for 43 per cent of the state budget. These consist of transfers to individuals regardless of their location. (Unemployment benefits are not included in this cluster but instead are allocated to the first cluster.) Hence, pensions, sickness benefits and family income support programs are the core of this cluster. Differences among municipalities in terms of transfers per capita are expected to be comparatively small, largely reflecting demographic variation.

9.4.2 Regional Dependence on Public Resources

Between 1985 and 1991, total transfers in Sweden grew by almost 33 per cent from 424 to 563 billion SEK (constant 1985 values). Figure 9.3 presents detail on the distribution of total transfers to a set of ten Employment (E) Zones for 1985 and 1991. This is a regional classification of ten clusters of functional employment areas, each representing a relatively stable and homogeneous set of employment areas (see Table 9.3).[6]

Sparsely populated areas with dominant regional centers (SPA) are the set of areas traditionally deemed problematic and targeted by direct investment by central government. These are located mostly in the northwest with a few small areas in the south. Stockholm, Gothenburg and Malmö, Sweden's three largest cities, each have their own classification in this schema. The Regional Service Centers Group (RSC) covers fairly self-contained and diversified regions in terms of production. Almost all RSCs have a core region with higher education facilities and regional administration. These areas cover most of Sweden's industrial zones. Small/medium size and differentiated areas (SMD) are spread throughout Sweden and represent diverse regions characterized by a number of different small to medium sized set of industries. Large

---

[6]The actual process involved the aggregation of Sweden's 284 municipalities into 111 labor market areas with largely internal commuting. A cluster analysis was performed using four sets of fundamental grouping characteristics: 1) size of land area, population, and proximity to urban zones, 2) educational standards, 3) industrial structure, and 4) structure of company size. This analysis yielded a set of ten readily classified regional "types".

scale industry (LSI), small scale industry (SMI), rural (RUR) and one company dominated (OCD) areas are self-explanatory.

Table 9.3. Definitions of E-Regions

Employment Zone Name and Description
Stockholm
Gothenburg
Malmö
Regional Service Centers (RSC)
Sparsely populated with a dominant regional center (SPA)
Small/Medium sized and Differentiated (SMD)
Large Scale Industry (LSI)
Small Scale Industry (SMI)
Rural (RUR)
One Company Dominated (OCD)

Source: Carlsson et al. 1993

Stockholm and the regional service center areas receive the lion's share of payments, mostly due to their large populations (Figure 9.3). It can be argued that, in a national context, payments to peripheral regions – such as SPAs – are of minor budgetary importance. The figures for 1991 show that total funds are not spent very differently than they were in 1985. Overall, these figures suggest that there is a certain amount of stability built into the complexity of the transfer system.

There has been large growth in per capita total transfers across all regions. What stands out from these data is the large per capita transfer value associated with sparsely populated areas with dominant centers. As mentioned, these are the traditionally problematic areas which have been targeted with large amounts of funds. Figure 9.3 also makes clear the large and growing values for other rural regions.

Growth in per capita transfers between 1985 and 1991 has occurred mostly for large scale industry regions with the bulk of this growth from increased expenditures for acute local problems. These areas have been quite adversely affected over the past few years and much intervention into the local labor markets has been evident. Although the more urbanized zones continue to receive growing amounts of transfer payments, Stockholm stands out as having relatively higher growth.

This arises mostly from increased expenditures in response to local labor market problems. Sparsely populated areas with regional centers evidence lower growth partly because the value in 1985 was already high. What is clear is that there is relatively large growth in every region, most especially in the non-urban locations. The sources of this growth arise from the vast number of programs and policies in effect and contribute to regional growth rates in a number of different patterns.

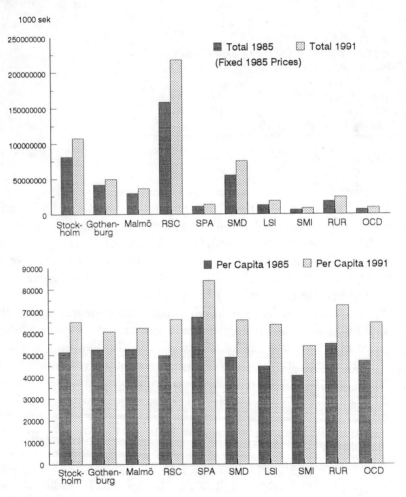

Figure 9.3. Transfers to E-Zones: 1985 and 1991
Source: After NUTEK 1994:3 and SOU 1989:65

We can discuss major regional differences among these transfers using the categories under which they are made. For example, amounts associated with income transfers (pensions, sickness benefits, income supports, etc.) actually grew by less than the average for the total, and this category of transfer income was an especially important source of growth for one company dominated areas and sparsely populated areas with dominant centers. The service cluster grew quite strongly over this period, resulting from increased expenditures on administration, health and social care, agricultural support, and infrastructure maintenance. Growth in this cluster was especially important to explaining the increases for SMD regions. (See Figure 9.3.)

Not unexpectedly, expenditures associated with acute labor market problems grew faster over this period than any other category. Large scale industry areas and, to a lesser extent, rural areas, evidence clear increases. One company dominated areas and small scale industry areas also received large increases in response to various acute economic problems. Even the Stockholm area had a large increase. It seems that the rural areas relatively close to urban centers have not benefited from their location, but they show a substantial increase in terms of expenditures received. It is clear that by the early 1990s, all regions were facing acute social and economic problems which the state government tried to ameliorate through special programs and policies.

In summary, the regional distribution of total transfer payments has become more unequal over time. Prior regional policy, with its focus on longer-term regional development has been forced to move to react to the acute problems being faced by all regions. The several hundred programs that comprise the bundle of transfer activity in Sweden evidence major shifts across functional categories. Expenditures on research, education and culture, the gauge for long-range intervention, grew at less than the national average for all transfers. The acute local help transfers grew almost 60 per cent above national average. The structure of all of these payments, however, makes it a rather complicated task to disentangle just how they will impact upon particular regions.

9.4.3 Spatial Changes in Public-Sector Employment

In a geographical sense, the public sector is and continues to be important in contributing to increasing employment in most regions. Figure 9.4 presents information on public sector employment levels for 1980, 1985 and 1990 across the ten regional categories.

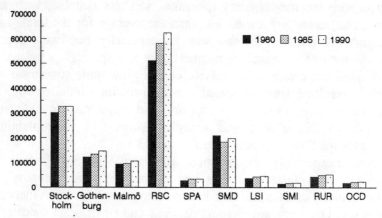

Figure 9.4. Public Employment by Ten E-Zones: 1980, 1985 and 1990
Source: Employment data, Statistics Sweden

At the national level, public employment between 1980 and 1990 was a large and growing part of total employment, increasing from about 1.4 million to approximately 1.6 million jobs over the period.[7] This represents an expansion from just under 35 per cent of total employment in 1980 to almost 37 per cent in 1990. By 1992, this figure had surpassed 40 per cent. The spatial distribution of public-sector employment is quite varied. All but one E-Zone evidences increases in levels of public sector employment both between 1980-85 and 1985-90. In terms of public-service employment levels, the growth is largest in the Regional Service Centers category which accounts for about more than 40 per cent of total national growth between 1980 and 1985 as well as between 1985 and 1990. It seems that Regional Service Centers in Sweden serve as important nodes in the geography of the welfare state.

---

[7]Information on employment in 1980, 1985 and 1990 is generated from a consistent set of data on employment by 5-digit SNI (ISIC) code and by municipality aggregated to appropriate levels. Employment data covering the 1989 to 1992 period are also used for separate analyses. These two data sources are not consistent and derive from a number of different sources.

Growth in public sector employment as a percentage of total regional employment also evidences wide regional variation. The heavy reliance of the Northwest regions (sparsely populated with a dominant regional center) on public-service employment is notable. In 1980, public employment in this region was about 37.5 per cent of total regional employment. Large additions between 1980–1985 increased this to just over 43 per cent and continued increases between 1985–1992 leave this region with almost 44 per cent of all employment in the public sector, the highest reliance on public-sector employment in all zones.

9.5 Summary

9.5.1 The National Scene

The history of the Swedish welfare model makes clear the prominent role played by the central government in Sweden since the 1930s. From its beginning, the Swedish model has focused on a number of distributional functions – from its inception as a means to provide guidance and administrative assistance for poor relief through municipal reforms geared at insuring the successful implementation of new social services such as schools and other social infrastructure. The balanced growth of the Swedish economy across its diverse regions has been shown to be of critical concern to the central government from the beginning.

The major public support necessary to underpin government's assumption of such a large role was and remained evident over several decades. Central governments' efforts were deemed appropriate and steady increases to these efforts were introduced over a relatively long time-span. The continued public support and stable growth of the welfare state was so firm that the economic boom of the 1960s provided powerful fuel for an even larger growth in government's role in a number of new and larger social programs. "The Record Years" marked the heyday of the Swedish model.

The diversity of the policies and programs can be seen from its very roots. The Swedish Welfare Model consists of a vast panoply of changing policies and programs which have evolved over time; sometimes by steady growth in well-defined areas, sometimes by rapid introduction of policies to address acute problems. At its core, however, is an ideological belief in the positive power of government to effect change and help redistribute resources in a more 'equitable' manner. Although companies have paid large amounts towards the expenses involved, the results have yielded fairly long periods of stable institutional settings and rules.

It is often the case that a number of approaches are used to solve certain problems. For example, labor market interventions run the gamut both in terms of aid to companies and aid to the worker. Companies have been given investment help, subsidies, tax relief, etc. Companies provide closing information, retraining, relocation. Workers are provided direct relief, retraining, relocation, etc. In essence, there is no 'one' solution sought nor any 'one' solution applied. Many programs, however, are introduced in a general form so as to strengthen the uniformity over the country.

An examination of the government's role in attempting to promote regional equity shows the diversity of approaches that can be taken. These run the entire range of investment in physical and social infrastructure to placement of government activities. Throughout the 1970s, some of the centralized government functions were deliberately moved out of Stockholm to other regional centers.

However, actual aid to regions represents a mere fraction of the transfer funds that result in final regional distributions. When these expenditures are synergetic with other programs, policies and events, and of great importance in a period of economic growth, it appears clear that regional equity is not horribly costly at the same time that it is socially beneficial. It is possible that the current state of the Swedish economy has broken the synergy that existed between stated and implemented regional policy and other transfer policies and programs. In fact current economic problems in Sweden are calling into question some of the fundamental ideological bases of the Swedish welfare model.

Economic decline has led to a new set of economic and social problems. Extant policies and programs, many devised to address specific historical regional issues, are being used much more to address acute problems across a wider range of regions. These shifting expenditure patterns, across both regions and sectors, are leading to increasing differences among regions. Growth in disposable income between 1989 and 1991 was above average in the urbanized areas of Sweden and below average everywhere else. We have already provided evidence of the large increases in per capita transfer payments being made to the more rural regions.

Strategic plans by the central government for regional equalization rest on the clusters of transfers that focus on long run change: namely research, education and culture, and, to some extent, the service cluster with its focus on infrastructure. Current problems are forcing the government into using tactics that rely much more heavily on reactionary and short term programs. It is clear that strategy is in conflict with current tactics.

The nature of the Swedish political system, with its focus on long-term policy implementations, make forecasting the future even more difficult. That the central government will continue to provide large amounts of transfers to regions is not in question. The traditional right of the government to impose restrictions on how those resources are spent, however, is changing. It is not merely the provision of funds, but the way those funds are used which has brought about the current levels of regional equity.

Recent moves have been made to decouple much of state funding from its end use. This will certainly result in greater diversity and difference between and among regions. Growth in transfer expenditures by central government will most likely have to decline. Determining how to share increases in an equitable way is much easier than deciding how to distribute cuts. This, however, is the task that needs to be done. The large growth in dependence on public sector funding and employment evidenced by a number of regions makes it clear that such cuts could greatly impact regional equity issues.

Maintaining regional equity goals during the period of cut-backs will appear more costly than ever. Studying and understanding more fully the actual impacts of transfers to regions will help make more explicit exactly what the impacts of these reductions will be both in terms of national efficiency and regional equity. One major issue that needs to be examined concerns the nature of the present and changing dependence on the welfare state in the various regions and the interrelations between the public and the private sector at the regional level.

9.5.2 Entering the European Scene

Increasing demand for international competitiveness, the challenges introduced by the EU-membership, and technological renewal of industry have all intensified Sweden's debate on regional development. Compared to the European average of 147 inhabitants per km^2, Sweden's 21 inhabitants per km² makes it one of the most sparsely populated nations in Europe. The economic and social functions of its peripheral areas are of major importance. The forest industry, the dominant net export industry, depends on domestic raw materials from these sparsely populated regions. More than 50 per cent of electric energy is produced by hydroelectric plants also located, to a large extent, in these regions. Yet those regions lack a dynamic future oriented industrial base. Employment structures rely on production of natural resources and on the public sector – the welfare economy that has evolved over the past three decades.

As production based on natural resources continued to demand less labor, the growing public sector was able to compensate for reductions in employment in many rural areas, at least up to the mid 1980s. Expansion in the health care sectors, day nurseries, etc., helped to ameliorate declining industrial employment. The failure to introduce more future oriented production/employment sectors (e.g., the expanding information sector) was hidden by the fact that a growing public sector kept levels of unemployment down.

In discussions preceding the Swedish referendum on EU membership in 1994, the debate on the consequences of European integration on the Swedish welfare model intensified and the spatial implications of reforms and reductions in the public sector were analyzed. In summary, these studies concluded that the specific and direct effects of European integration are expected to be limited. The primary reason is that cutbacks in public expenditures are largely necessary with or without EU membership. Increased productivity in the service sectors has been realized and is already stimulating reorganization of municipal services.

As the combined result of a political shift towards market solutions and a confronting of budgetary problems (and to a lesser extent adjustments to the EU rules), substantial reductions in the state budget are anticipated (Tapper 1992). One estimate reveals amounts of approximately 11,000 SEK per capita per annum in peripheral regions and 8,000 SEK per capita in intermediate zones, with central regions benefiting from tax reductions and limited contraction of public services. The primary reason for the larger negative impact in the periphery is due to the present high dependence on public services and transfers coupled with relatively low income levels. Thus tax reductions will have relatively small positive effect on buying power in these regions. The figures are hypothetical, however, and should only be looked at as examples of what would happen if proposed budget reductions were carried through.

Nevertheless, this example indicates that a spatial perspective on changing macro conditions is needed. Redistributional effects of policy changes need to be taken into consideration as well as the needs and potentials of the different rural areas. Such a policy can no longer be restricted to one or two sectors, as agricultural and traditional regional policy; it must include all sectors in a strategic and progressive package. The major task would be to facilitate the move for rural areas to enter the growing sectors of the economy (cf. Persson 1993; 1994).

The European integration process has quickly changed the way that regional differences are perceived in Sweden. Only a few years ago the metropolitan regions of Stockholm, Gothenburg and Malmö were viewed as the centre, and the north of Sweden was the periphery which demanded special attentions so as not to lag behind. Today it is a common view that these metropolitan regions themselves are, in effect, a European periphery demanding infrastructure investments to keep pace with the growing regions in Europe. In this perspective, the north and rural parts of Sweden can only be helped if other regions in Sweden successfully compete in the larger European scene.

This debate, which cannot even conclude whether Stockholm should be viewed as periphery or centre in a national policy, reflects the present confusion about the future of regional and rural policies in Sweden. While the EU is paying closer attention to the regional dimensions of central policy, this is not yet very obvious in the state policy in Sweden. More and more, regions themselves are increasing their own activities in an attempt to find a new direction as the leading role of the state appears to be lessening (Persson & Westholm 1993; 1994).

References

Alber, J. (1983), 'Some causes of social security expenditure development in Western Europe 1949-77', In Loney, M., Boswell, D. and Clarke, J. (eds), *Social Policy and Social Welfare*, Open University Press, Milton Keynes.

Axelsson, S., S. Berglund & L.O. Persson (1994), *Det tudelade kunskapssamhället*, ERU-rapport 81, Stockholm.

Berglund, S. & L.O. Persson (1994), 'Towards a polarization of the urban labor market in Sweden', Paper presented at the Conference Cities, Enterprises and Society at the

Eve of the XXIst Century, Lille, March 17–18, Research Group on Regional Analysis (FORA), Royal Institute of Technology, Stockholm (mimeo).

Carlsson, F., M. Johansson, L.O. Persson & B. Tegsjö (1993), *Creating labour market areas and employment zones*, CERUM report, Umeå.

Castells, M. (1989), *The informational city*, Basil Blackwell, London.

Ds (Ministry report) 1994:61, *1993 års redovisning av utvecklingen i kommuner och landsting*, Civildepartementet, Stockholm.

Edin, P.A. & B. Holmlund (1994), *Arbetslösheten och arbetsmarknadernas funktionssätt*, Bilaga 8 till Långtidsutredningen 94, Finansdepartementet, Stockholm.

Eklund, K. (1994), *The modernization of the Scandinavian welfare state*, RURE Newsletter, No. 6, Department of Geography, Uppsala.

Esping-Andersen, G. (1990), *Three Worlds of Welfare Capitalism*, Policy Press, Cambridge.

Fördelningseffekter av offentliga tjänster, Ds (Ministry report) 1994:86, Rapport till expertgruppen för studier i offentlig ekonomi, Finansdepartementet, Stockholm.

Ginsburg, N. (1992), *Divisions of welfare. A critical introduction to comparative social policy*, Sage, London.

Holmberg, S. (1993), SOM-rapport 11. Statsvetenskapliga institutionen, Göteborgs universitet, Göteborg.

Myhrman, J. (1994), *Hur Sverige blev rikt*, SNS-förlag, Stockholm.

NUTEK (1994), *Sveriges ekonomiska geografi, Regioner i ömsesidigt beroende*, Närings- och teknikutvecklingsverket, Biliga 5 till Långtidsutredningen 95, Finansdepartementet, Stockholm.

NUTEK (1994:3), *Statsbudgetens regionala fördelning*, Närings- och teknikutvecklingsverket, NUTEK Förlag, Stockholm.

OECD (1994a), *Economic report: Sweden*, Paris.

OECD (1994b), *National Revenue Statistics*, Paris.

Persson, L.O. & U. Wiberg (1995), *Microregional fragmentation; contrasts between a market economy and a welfare state*. Physica-Verlag, Heidelberg.

Persson, L.O. & E. Westholm (1993), 'Turmoil in the Welfare System Reshapes Rural Sweden', *Journal of Rural Studies*, 9, 4, 397–404.

Persson, L.O. & E. Westholm (1994), 'Towards the new mosaic of rural regions', *European Review of Agricultural Economics*, 21–3/4, 409–427.

Persson, L.O. (1993), 'Peripheral Labour Markets under the pressure of Change', In Gilg, A.W. (ed.), *Progress in Rural Policy and Planning*, 3, Belhaven Press, London.

Persson, L.O. (1994), 'Social Structure and Processes of Change in Marginal Regions', In Wiberg, U. (ed.), *Marginal Areas in Developed Countries*, CERUM Report, Umeå.

Regeringens proposition (Government Bill) 1993/94:40 (1994), *Bygder och regioner i utveckling*, Regeringskansliet, Stockholm.

Social Rapport 1994:10 (1994), Socialstyrelsen, Stockholm.

Socialstyrelsen (1993), *Hemlösa i Sverige – en kartläggning*, Socialstyrelsen följer upp och utvärderar 1993:13, Socialstyrelsen, Stockholm.

Söderström, L. (1988), *Inkomstfördelnings och fördelnings politik*, SNS Förlag, Stockholm.
SOU (Statens offentliga utredningar) 1989:65, *Staten i geografin*, Arbetsmarknadsdepartementet, Stockholm.
SOU (Statens offentliga utredningar) 1992:19, *Långtidsutredningen 1992*, Finansdepartementet, Stockholm.
Tapper, H. (1992), *EG, statsbudgeten och den regionala balansen*, Ds (Ministry report) 1992:80, Arbetsmarknadsdepartementet, Stockholm.
Yearbook on Nordic Statistics 1994, Nord 1994:1, Copenhagen.

CHAPTER 10

Are Leaping Frogs Freezing?
Rural Peripheries in Competition

Jukka Oksa
University of Joensuu

10.1 Periphery, Competition, and Rurality

As new societal structures are being consolidated, they include their own set of core-periphery relationships. They might be different from the earlier ones but they also repeat features of the past. For those living in peripheries, periods of restructuring arouse new hopes. Ancient stories of leaping frogs are told again. These belittled creatures compete with stumbling giants and win by making amazing jumps. Earlier neglected peripheries become centres of the new order. That is why, in such times as ours, there is a need to update the concepts of periphery.

Peripherality in both a spatial and a social sense means being out of the centre of resources, such as wealth, power, or connections. The concept of rurality has a very similar relationship to its opposite: urbanism usually means trades, industries, power, learning, and progress. The developing part of today's world is often conceptualised as the urban system. In this discourse, rural is something residual, somewhat sub-, or only partially urban. If someone peripheral or non-urban is doing well in competition, (s)he is defined to be something exceptional, extraordinary, whose achievement needs a separate explanation, like a child-genius of music or mathematics. If a small sawmill town in the middle of forests becomes a significant centre for a special field of art, for example training of brass musicians, this phenomenon is easily portrayed as being somehow connected with periphery. At least it is celebrated as a heroic effort to cope with "The Peripheral Experience."

10.1.1 Innovations and Frog's Leaps

There also exists a more serious and academic tradition of discussing peripheries in terms of change and innovation. In the 1960s, Sahlins and Service (1960) wrote an influential book on the non-linear nature of social evolution and reminded of the privilege of historical backwardness. Johan Galtung (1966, 1968) even constructed a general theory of unbalanced ranks to account for peripheral behaviour. Historical geographers have discussed how former peripheries, because of having more innovative dynamics, have managed to take a frog's leap and have become centres of a new order (e.g., Dodgshon 1987, 14–16).

One particular feature should be recognised in this stream of writing. Innovative leap-frog phenomena are not as common in the remote periphery as in the semi-periphery, near the edge of the core. According to this reasoning, the semi-periphery does not have the rigid institutional burdens of the centre. Handicaps that can turn an established centre into a stumbling giant are, for example, out-of-date hierarchies of information, inflexible divisions of labour, illusions of everlasting success, and power structures suffocating internal criticism.

Recent debates about rural strengths have somewhat similar emphases. Just as the defenders of the innovating periphery regard the core as rigid and un-inventive, the friends of the countryside tend to underline the urban restlessness (rush and anguish), immorality (harshness, indifference, crime) and unnaturalness (unhealthy environment, pollution, congestion). The new possibilities of the countryside are seen in the resources that the town is lacking, namely, the nature and the cultural legacy of the traditional past. These may become attributes of products, like healthy food or rustic handicraft, of attractive housing districts, cosy telework centres or tourist resorts, or of an idyllic image in general. The natural is often linked with ecology and health. These qualities are something for which the consumers of the city are ready to pay. In this manner, rural, which was yesterday regarded as something backward and stale, can be associated with good and vivid life.

10.1.2 Peripheries in the Post-modern

History is giving us more thorough lessons than only the redistribution of existing central and peripheral positions. The issue is not only who is occupying which position, but also the number and quality of positions.

A significant change is the increase in the number of dimensions, hierarchies or pyramids (whatever you want to call them), in which one can either act as a core or be left in a periphery. Also, the pace of changes has become more rapid, which means that a position as the centre of some activity can last a shorter or longer time: a decade as a world centre of men's fashion, a year as the Cultural Capital of Europe, a fortnight as the scene of the Olympic Games, or one week as an organiser of a successful summer event. Soon every place can be a centre of something at least for a short period of time.

Does this mean that the talk about cores and peripheries becomes a reminiscence of the old-time stable societies? Has this post-modern mesh of flexing relationships diluted the meaning of the concept of periphery? The hierarchies of spaces have not dissolved into thin air, although they have become more complicated and refined. The sense of peripherality can re-emerge in more concrete dimensions and historical settings. For example, if the importance of knowledge creation and information processing is increasing, then university and service centres would increase their status. If the small sawmill town of our earlier example had a university, and if that institution were generating relevant knowledge for the manufacturing and consumption of wood-based products, then the town might become a valuable node in the international forest sector network. For the future of that town, this position might be more fundamental than the earlier mentioned position as a centre of training brass players. Naturally one might also argue for another order of priority. Although reductionist simplification should be avoided, reassessment of the significance of different dimensions of core-periphery relations is needed now, as the process of European integration is progressing and consolidating a new set of core-periphery relationships.

10.2 The Changing Peripheral Role of Rural Areas of Finland

In this section, the historical change of Finland's rural areas will be discussed from the viewpoint of core-periphery relationships, in which economic and political factors are the most critical determinants of peripherality. The focus will be on the changing economic role of the countryside, the form of marginalisation, the mode of local responses, the mode of regulation (or integration) of the rural areas, and the power coalitions connected with these policies. The empirical observations primarily concern Finland's eastern rural periphery. Finland can be regarded in many respects as a typical Nordic Welfare State.

Finland's eastern periphery belongs to the sparsely populated north of Europe. If one wants to emphasise its natural attractions, it can be described as a lake and forest district. If wilderness is preferred, it can be described also as a part of the northern Taiga. In history its population has paid taxes to the Swedish Crown, the Russian Emperor and the Finnish Republic. In today's European map it comprises a part of the border between the European Union and the Russian Federation. Its forests have been the wooden foundation of Finland's forestry multinationals. To be able to understand the current contradictions, it is necessary to review shortly the phases through which the present structures of this rural periphery have been constructed.

10.2.1 Rural Territory as the Frontier for Expansion

For a long period of history, the northern rural space was a frontier of natural resources. It offered land for settlement and raw materials and energy for industries. Agriculture and forestry were the bases of both the private economy and the State economy. A specificity of Finland's development, compared to other Nordic countries, has been the scope of private, peasant forest ownership. Finland has been characterised both as "a peasant state" and "a forest sector society". The Finnish nationalistic movement at the turn of the century was a coalition of cultural intelligentsia and peasants, which were united by the Finnish language. The core of the nation's political structure and national coherence was built of the compromises of the two wings of the forest sector elite, private forest owners (peasants) and forest industries. Both of them benefited from the export of forestry products, which defined the role of Finland between the two world wars in the international division of labour (Koskinen 1985).

The land issue was an important mobilizer of political movements, and land reforms were the key policy tool used to build cohesion across the social divide and to solve societal conflicts. After Finland's independence (1917) and the violent class war of 1918 the central mechanism for creating new social integration was land reform. Private land ownership was a cement of emerging rural coherence in the 1920s. Settlement programs, which resulted in the creation of new farms, were used to heal the wounds of the Second World War, when Karelian evacuees and war veterans were resettled. Over 400 000 Karelians moved away from those parts of Finland that were annexed to Soviet Union after the war. As most of them were farmers, their losses were compensated by receiving new land elsewhere in Finland. Thus, clearing new agricultural land was continued in Finland until the late 1950s.

10.2.2 Rural Population as the Labour Reserve for Industrial Growth

Industrialisation and urbanisation took place in Finland relatively late, but very rapidly. In the 1950s the Finland State policies started emphasising industrialisation and in the early 1960s there was a political turn towards industrial growth and adaptation to the international division of labour, first in the EFTA co-operation, and later in the free trade agreement with the EEC.

Profound rationalisation took place in the organisational framework of both farming and forestry. In agricultural policies there was a move away from subsidising small farms towards improving productivity, specialisation and mechanisation of viable family farms. Differentiation took place both inside villages and between villages.

In the forest sector there were campaigns for more effective production of roundwood. A re-organisation of work along with new cutting and transport technologies were introduced to forestry. These had dramatic impacts on the fabric of rural communities especially in remote areas. New salaried forest workers replaced the old seasonal lumberjack and his horse. The entire local form of rural economy based on small-farming and winter-time forest work was crushed within ten years.

The rapid structural change turned into a large wave of migration from peripheral areas to industrialising centres during 1965-75. Later this was named the "Great Move". The inhabitants of villages were frustrated. They felt that the society was forgetting them. They reacted by giving their support to a protest expressed by the Finnish Rural Party (SMP). In the Parliament elections of 1970 and 1972, this movement surprised all of the political analysts by getting 18 members (of 200) into the Finnish parliament, the greatest loser being the Centre Party (earlier the Agrarian Party). This protest gave impetus to the public debate on the issues of small-farming and less developed areas. The restructuring of the 1970s was accompanied by intensive political struggles and several reforms of regional, employment and social policies.

10.2.3 Rural Communities as Partners in the Welfare State

The political struggles of the 1970s produced a Finnish version of the Nordic Welfare State. It was not exactly the kind of industrial state that the proponents of industrialisation and urbanisation had had in mind. Some elements of the earlier peasant state or forest sector society continued their influence in the new context. The old wings of the forest sector elite, the private forest owning peasant and the forest export industries, still had their impact, but they had to share their power with the new actors of the political elite, the leaders of industrial labour organisations. In the implementation of the industrial growth policies the social democratic leaders were in influential positions. The Finnish Welfare State was for a long time run by the governments of the Social Democrats and the Centre Party. In this coalition the Centre Party was able to act as representative of the farmers and of the industries.

The strong agricultural and forestry bloc managed to give its own character to the welfare system, through paid holidays, national pension plans, child allowances, health services, and guaranteeing public services to the rural population. The peasant lobby geared the agricultural policy to supporting a model of modernised one-family farms effectively producing raw material for food industries. In eastern Finland these products were milk and meat. The development of farming income was guaranteed by central income policy agreements.

Regional policies can be regarded as one set of compromises between industrialisation coalitions and the rural bloc. Finnish regional policies were quite similar to those of many west-European countries. During the 1970s state supported industrialisation of rural centres was practised. Investments in both capital-intensive forest export industries (pulp and paper) and labour-intensive light manufacturing (wood manufacturing, metal products, clothing) were subsidised by the state. In the 1980s regional industrialisation policies were found to be ineffective, and new regional policy measures were geared to support the establishment of small and medium-sized firms (product development, marketing, etc.).

The most substantial impact upon rural development, however, was not brought about by regional policy measures but by general welfare reforms. Those had a balancing effect on regional disparities in the 1980s. Several reforms, such as a comprehensive school system, health centres, and social services added to the functions of municipalities, which created new jobs, often offering employment to women. Many rural municipal centres grew rapidly and captured the migration outflow from remote villages. In addition, the rural political response developed more consenting channels and returned to support governing political parties and national interest organisations.

A new institution of village-level coherence was created, promoted and successfully spread throughout the country. The village committees sought to unite various social groups inside the village. The modernised village has a population consisting of several professional groups. It can be called a localised miniature of the industrial welfare society. The village committee may co-ordinate several organisations of the village, such as the forest worker's union, the agricultural producers' association, the local shopkeepers, the bank office, the hunting association, etc. Typical activities of a village committee are, for example, clearing sports field or village beaches, organising summer festivals and skiing or fishing competitions. They often make proposals to the municipal administration, for instance to improve local services, or to install electric lighting on public road or a skiing and jogging tracks.

10.3 The Impacts of the Welfare State in the Rural Periphery

The Nordic model of the Welfare State has had its own rules of marginalization. Those social groups which did not have any resources for social mobility (education, skills) were marginalized. They were pushed out of the labour force, or excluded from policy-making and left out of the scope of reforms. Remote and rural location tended to reinforce these exclusion mechanisms. Rural people tried to cope with them by economic and political means. It seems that gathering the forces, fighting for reforms, and making investments in the 1970s, brought some positive results. The 1980s was a more restful period for rural areas, and some impacts of national economic growth were channelled to the countryside. During the 1980s Finland's rural municipalities were able to maintain their population, although inside the municipalities the move toward the municipal centre continued.

During the two decades of the Welfare State, specific economic structures were established in rural areas: specialised and mechanised family farming, efficient highly mechanised forestry, sawmills, and manufacturing plants, and an extensive network of public services. Many of the general features of the Nordic Welfare State have reached the rural communities. The significance of kin as a source of personal security has diminished. The family has shrunk into a nuclear family, or a household of a single or a couple. The result is a kind of state-supported individualism, with readiness to rapid social and geographical mobility.

Other attributes of the Nordic Welfare State are a developed system of public services and a large share of women working full-time in the local labour market. Also, the change of the countryside is remarkable. In a typical rural municipality, over half of the economically active population works in services, and half of them in public services. Agriculture and forestry make up about one third of the employed. One must remember, however, that the bulk of the service jobs are located in the municipal centre, not in the sparsely populated remote areas. But some impacts are also reflected there. Although the combination of a female cattle-farmer and a male part-year forest worker had been destroyed by mechanisation of forestry, some small-farmers have survived by combining male farming and female service work (see Sireni 1994).

An important effect of the welfare state has been the arrival of a newly educated stratum of service professionals to rural centres. In addition to developing their own professions, they are important in defining and expressing rural interests. They also represent new values (consumerism, environmentalism), which sometimes clash with the more traditional values of land-owners and agricultural producers. New professionals provide the backbone of the local expertise in rural planning and politics.

10.4 Making of Rural Peripheries of the European Union

Presently, the rural periphery of Finland is attempting to find a new role at the fringe of the European Union. This process is made up of contradictory elements, some of them resembling the beginnings of the industrial growth project in the 1960s. At that time the social consequences of rapid economic restructuring were neglected, until the political protests forced the ruling coalition to accept rural elements into their reforms.

Groups that see their future prospects connected with European markets, have been mobilising to prepare for Finland's membership in the European Union. The political turn from the Welfare State to the new project of European Union membership was taken stepwise. First a red and blue government of the Right and Social Democrats started economic structural change, liberalisation of banking and financing, etc. This government represented interests of export industries and internationally oriented urban service classes. The policies of this coalition, however, created a political backlash, especially in the rural parts of the country. The protest votes in the election of 1991 went to the Centre Party, which had been in opposition. After the election this party formed a government together with the right wing parties, leaving the Social Democrats in opposition. This government took the final and decisive steps towards the EU membership. Its push for rapid economic structural changes favouring export industries and cutting public spending, ended up with an unemployment rate of over 20 percent. The compromises of the rightist Coalition Party and the Centre in the government, and the support by the Social Democrats for the integration in the opposition, have steered Finland into the European Union.

There are two intertwined processes in today's change. One is the restructuring process of adapting Finland's economy to the European markets. This process was actually going forward quite briskly in the late 1980s during the economic growth period. Later it has been interwoven with the process of recession of the economy and the budgetary crises of the Welfare State. Together these create a gloomy perspective for the rural periphery.

In the middle of the European Union integration process there has been a turn in Finnish rural policies. The comprehensive new rural policy programme did promise to support the competitiveness of rural areas and to create new rural industries. These promises have faded into a new wave of differentiation of the countryside. Particularly the paths of the countryside near the towns and of the remote countryside are going in opposite directions. The special needs and problems of remote areas are being neglected. Partly this is due to the weakening of the Welfare State and the elimination of those elements inside the new rural policy that were developing new models of social services in sparsely populated countryside. Partly this is due to the focus of national rural interest representation on issues of agricultural policies.

Rural communities have found it hard to see positive prospects in the integration project. It has been associated with harsher competition for small industries and a forecasted halving of the income of farmers. The cuts in public budgets are threatening the public sector jobs of rural centres. New market oriented ways of avoiding marginalisation are hard to find. These problems were reflected in Finland's 1994 referendum on EU membership, in which the majority of rural voters said No. The opponents almost succeeded in postponing the membership by getting 43 percent of the total votes. The question of forsaken and marginalized rural periphery will be rising on the political agenda again.

10.4.1 Promise of Integration Through New Rural Policy

In the late 1980s, during the growth of Finland's economy, possibilities for successful rural restructuring were seen positively. A public tool for aiding this change was the new rural policy. Finland's National Rural Programme aims to reallocate public resources to support all sectors of rural economy, instead of agriculture and forestry alone. The main objectives of the programme are: to promote rural networks and rural competitiveness, to focus resources on the expanding rural industries, to reorganise the provision of services, and to support pluriactive farming. (See Rural Policy Committee 1992, and also Malinen et al. 1993.) These same policy lines are encouraged by international recommendations (Commission of the European Communities 1988; OECD 1993).

The new rural policy aims to diversify the rural economy in order to make it more competitive. The objective is to be served by several changes or shifts of emphasis in rural development policies. First, a shift from agricultural rurality to housing rurality underlines the need to attract new inhabitants who are working also in towns. Secondly, the contradictory measures of administrative sectors are to be co-ordinated. Thirdly, the responsibility for the provision and co-ordination of public services is to be transferred to local municipalities. The new rural development programmes emphasise the role of local partners in planning and implementation of projects. In the new rural policy programme, Finnish rural areas are divided into three development zones: the vicinity of towns, the countryside proper, and remote rural areas. Their different priorities and problems should be taken into account in policy.

10.4.2 Excursion to Rural Telematic Projects

One example of the policy strategies of the new rural vision is the promotion of telematic systems. The existing experience of almost one decade of these projects reveals both weaknesses and possibilities of new development models in the countryside. The telematic experiments have an important position in new rural policy documents. This is understandable because new networking technology is regarded as one of the key areas for the future. Their applications can reduce the role of geographical distance. The EC programme of telematic systems (1990-94) included a specific area of telematic systems for rural areas (the so called ORA-programme). Also a new rural-policy report by OECD (1993) includes measures for reducing the isolation of places by modern communication systems. The EC programme mentions the Scandinavian model of community teleservice centres, often called telecottages, as an interesting experiment in this field.

The new information technologies have also been envisioned in Finland as a new device for the countryside to fight against isolation, to acquire jobs and to generate income, to attract educated people, and to sell rural goods and services. During the economic upswing the rural areas were seen as a future space for service activities. Information professionals were to live in rural communities, doing home-based work, setting up small enterprises and studios, improving their qualifications in on-line universities, participating in electronic conferencing systems, thus combining work and leisure into new community and family centred life styles.

The first telecottage, as far as Nordic countries are concerned, was founded by a Dane in northern Sweden in 1985. Five years later there were about 60 telecottages in the four Nordic countries. (Cronberg et al. 1990; Qvortrup 1990.) From Denmark and Sweden, the idea spread rapidly through personal contacts to eastern and northern Finland. The first two telecottages in Finland were founded in 1986, and their total number increased to over forty by 1992.

The typical telecottage has a slightly different profile in the different Nordic countries. According to Cronberg et al. (1990) the Swedish model emphasised home-based work (decentralisation of public jobs), the Danish model has been farmer-oriented, and the Finnish model has focused on education. Nevertheless, the definition of the telecottage has three common elements: new information technology, openness to local people, usually a rural community, and provision of guidance in the use of technology.

The cases of the two first Finnish telecottage localities in the easternmost province of Finland, North Karelia, illustrate the conditions of rural telematic experiments in different types of rural areas. The case of the Ruvaslahti village telecottage portrays the visions and changing realities of telematic experiments in a remote village. The experience of the Kontiolahti municipal development strategy is an example of a rural area in the vicinity of a university town.

The Information Society Comes to the Remote Village

From the beginning the Ruvaslahti telecottage has rested on the local cooperation of three groups: the experts that came to the village from outside, the village committee representing the local inhabitants, and the continuing education centre which represents the municipality and takes care of the administrative routines. This local triangle has been supported financially by regional and national government. The governor's strategic plan for North Karelia included the notion of the province as an experimental area of rural information technology.

Although efforts to utilise computers in local farming brought about disappointments, the telecottage has kept running. After the subsidised period the telecottage had to finance its activity independently. The role of the telecottage has been twofold. One is producing services to the village, and the other is creating income. The telecottage helps village organisations to run their routines and to formulate their proposals and initiatives.

The telecottage has maintained two permanent jobs in the village. The significance of these educated specialists goes far beyond the numbers. The Ruvaslahti telecottage has sold its accumulated expertise in rural development to the public sector. It has produced reports on telecottages and telework, a bibliography of rural research, created a regional database of rural experts, etc. The largest job was processing part of the national agricultural census data of 1990, which gave five part-time jobs for two months. The selling of services has been based partly on the available part-time female labour in the village (farmer's wives with, for example, commercial school training). In 1993 the Ruvaslahti telecottage made an agreement with the Central Association of Finnish Municipalities to function as its Village Information Service Centre.

The telecottage is not the only possible way of organising rural telematics. Many ideas have found other organisational forms. Other important types of telematic projects are home-based work, experiments to develop social work in rural areas, and experiments in education. Most of the projects are still local, small and cheap by international standards. However, among the newer projects there are also several nation-wide projects aimed at larger user groups. The projects are usually financed jointly by Finnish ministries and a regional government. In a typical case, the Finnish TeleCom has contributed the hardware, software, training and connection to telematic services, such as electronic mail, bulletin boards and access to public service networks. Training institutes and universities often provide the staff and expertise. The list of partners seldom includes any international organisations, private enterprises, or private foundations. The users of the projects are from the public sector (administration, education, social work) and agriculture. They very seldom come from manufacturing or private service sectors. However, an important feature is that the projects aim to create new links between different sectors. They try to combine activities, for example, services and agriculture or public services and private services.

It can be concluded that rural telematic experiments have been financed by regional and national government and also supported by public demand for services. They are deeply rooted in the reform traditions of the Welfare State, but they go beyond the framework of the earlier rural production structures.

The Information-Technology Strategy of the Municipality of Kontiolahti

Finland has a long tradition of municipal autonomy, and municipalities are still the most important actors in local level development policies. The new village committees are important representatives of village inhabitants but they rely on voluntary activity, and they have no economic resources of their own. The municipalities have a right to collect taxes, have elected decision-making bodies and have a permanent administrative apparatus. The new rural policies, as well as the Welfare State, have relied on the municipalities for the implementation of their programmes. The new rural policies, in addition, expect that municipalities take local initiatives and contribute to policies by defining development strategies of their own.

The municipality of Kontiolahti is an example of a rural area that is located in the vicinity of a province centre. Its development strategy has exploited the vicinity of Joensuu and its university. It has launched several development projects of information technology, constructed its infrastructure, raised the level of awareness of information technology, and has also built areas for housing the new middle class. It is not a model for peripheral municipalities in general. It is an example of a successful rural development strategy in the neighbourhood of province capital and university town. In addition, it is an example of successful image-building. According to its image, Kontiolahti is a dynamic community of the information age, with attractive rural surroundings, offering a combination of wild backwoods, an excellent golf course and high technology labour skills.

It is interesting to compare the situation of this information technology-oriented municipality with the earlier example of a more remote telecottage. Although the Ruvaslahti village telecottage received favourable publicity and fame, the Polvijärvi municipality did not use it as part of its development strategy or for attracting enterprises. The municipal development strategy was focused on industrialisation (manufacturing). The Ruvaslahti village was an exception in its attempt to bypass the municipal industrialisation policy and to link to the wave of information technology.

Why has the municipality of Kontiolahti been able to design and to maintain an information technology strategy? This is linked with another question: Which social groups have interests to promote information technology projects? Who tend to see their own future in terms of the new information technology? In a case study of a rural communications project in Kontiolahti, three such social groups were found: rural developers, rural service class, and rural entrepreneurs (Oksa 1993, 1994).

The rural developers are usually trained professionals, whose work or voluntary activism (political, interest groups) is connected with defining and expressing the "rural interest". These people can be politicians, planners, activists of rural based organisations, development project workers etc. They are attracted by the visionary promises for the rural future.

The rural service class could also be called "new rural residents" or "rural yuppies". They are well educated middle class fractions living in the countryside, often working in information professions. They want to combine the best of urban and rural life-styles, to live in the natural environment, to bring up their children in small communities, etc. They also want to have a successful career and utilise urban services, culture and information networks. For them, information technology gives a perspective of uniting urban connections and rural surroundings.

Rural entrepreneurs need better communications to develop their production and trading connections. They want to solve their practical problems with reliable technology that is easy to use. If it is available for a reasonable price, they employ it. The rapid adoption of telefax is a good example.

These three groups may form a coalition that promotes rural information technology. If these groups are strong, as they are in case of Kontiolahti, the result can be a long-term local development strategy. If these groups are weak, or their co-operation fails, information technology projects receive less attention and remain single experiments with no broader or longer impact.

This idea of local development blocs can also be applied to other new activities besides information technologies. Such blocs can be witnessed in issues such as industrial development, environmental projects, new food production and tourism. These blocs are designing, partly spontaneously, partly intentionally, new competitive strategies for villages and municipalities. An important common feature of these local development blocs and strategies is their orientation towards more intensive rural-urban interaction, which they use to fight the exclusion from the markets. At the same time as they are fighting against rural marginalisation, they are attempting to adapt rural areas to the demands of the new national policy project of European integration. Although they express solely rural interests, urban connections are their most important resources.

In these local development blocs the role of new rural middle-class fractions has been described as constructive. In the industrialised countries these social groups have changed the vicinity of metropolitan areas in massive scale. However, a group of British rural researchers describe these social groups as conservative, opposing new economic development projects, and defending their housing areas against changes (Marsden et al. 1993).

10.4.3 Recession and Marginalisation

Although in the 1980s Finland's rural municipalities were able to maintain their total population, this observation has to be qualified in two respects. First, the outflow of the young has continued from small and remote villages to the municipal service centres. Secondly, the number of employed persons had already decreased substantially by the late 1980s. This decline has developed into a drastic drop, during the public sector cuts of recent years. In spite of the new rural policy efforts, the labour market in the rural areas has been shrinking and the only locations with new job opportunities for the young are the provincial centres and their surroundings.

In the beginning of the 1990s, the economic recession and cuts of welfare state budgets have cast dark shadows on the visions of the service producing countryside. The self-government of rural municipalities has not been able to respond to the challenges of public cuts. The responsibility of planning and implementing the cuts was transferred to the bureaucracy or to private consultants. When the local experience is excluded, the only role left for local people is the one of protesting troublemaker (see Pyy 1994).

The public budget cuts also interfered with experiments in new rural policy. For example, there were projects to develop rural post offices into service points of new information networks, but these plans lost their ground when the bulk of rural post offices was closed. There are several programmes aiming at new ways of providing basic services in sparsely populated areas. One idea is that school kitchens could deliver warm meals to the village elderly needing such aid. This concept is hard to implement when village schools are being closed. In addition, the creation of private social welfare services threatens the rural approach to neighbourly help and care.

In remote locations with sparse populations, the town model of a local service enterprise is not viable, because of the small local demand. In such conditions a different logic of labour division is needed. A promising idea could be a multi-functional service point combining some welfare functions, voluntary activities and private services. "Unholy alliances" going beyond the traditional spheres of co-operation need to be allowed and encouraged, as they have been in some experiments of the new rural policy programme.

10.5 Political Conclusion: Freezing Frogs Don't Leap

Some groups of rural people have responded to the threats of marginalization by social mobility, protests and publicly stated demands for social reforms. In the beginning of this century the landless peasants were demanding land. In the industrial welfare state better life was pursued by occupational and geographic mobility, by political protests and social reforms. At the moment Finnish rural areas are again living through a transition period. The various conditions and forms of European integration are being contemplated and negotiated. In earlier decades the core forces of the peasant state were able to influence the policies of its follower, the welfare state. Now we shall see, if some elements of welfare state, in turn, are being transferred to the new social order emerging through the European Union? This will depend on the balance of forces, and on coalitions and compromising skills of political groups.

It is important to note the way in which the Centre Party has been using its influential position as the representative of rural interests. In the negotiations concerning membership in the EU the voice of the old elites of the forest sector society was strong. An ex-leader of the farmers' union took a leading role in the negotiations as the foreign minister, and dramatic public politicising was done around the issue of agricultural supports. This, once again, demonstrated the organising, mobilising and negotiating skills of the farmers' interest organisation. The issues of the new rural activities were set aside and all the attention was focused on the defence of farming interests.

The issue of a marginalized rural periphery is becoming critical again. In the new order of Europe people of rural peripheries are very likely to be left out of labour and product markets. However the reason is not exactly geographical. The bases of today's and tomorrow's marginalization is not kilometres, but poor network connections, and less participation in the important development streams of society. The connections depend more and more on the construction of information infrastructure, networks of education, research and development, and distribution chains to consumers. The growth of new activities and the emergence of new social groups in rural areas will depend on these connections. There are special risks that remote rural areas will be left out and that they will be, consequently, out of sight of the new social groups interested in rural life.

The agricultural support policies that were formed during the welfare state period have received an extension in the common agricultural policy of the European Union. However, it affects only a small part of the rural population, the rapidly declining farming population. In Finland the size of farms will further increase, which means a new wave of concentrated farm resources to fewer owners and fewer locations. In addition, in the future, the role of common European agricultural policy will certainly be redefined.

The new rural policy aims to diversify the rural economy and to make it more competitive. The shifts leading to this objective are taking form in a selective way. A shift from agriculture to housing rurality will favour locations within commuting distance to towns and with attractive surroundings. Transferring the responsibility of public services to the local level and shrinking state welfare budgets, have weakened the financial position of small rural municipalities, many of which are fighting for survival. Refocusing public rural development funds into new growing industries depends very much on local initiatives. This has slowed down change, especially in remote locations. The new regional development programmes need active local partners. For example, the famous EU-programme "Leader" is relying on local partnerships and local commitment. For local innovations, commitments and contributions by some form of local development blocs, or corresponding collaborative efforts, would be needed. The backbone of these blocs has consisted of service professionals, in remote areas, working in the public sector. The integration of villages into new information technology networks, for example by means of telecottage experiments, has relied on direct public funding, and, more recently, on public demand for services.

It is easy to see that the new rural policy has been in many ways embedded in the structures of the welfare state. As the resources of welfare state decline, the already apparent tendency that the impacts of new rural policy are restricted to the vicinity of towns, is enforced. If we look at the list of the new development processes of the countryside, every one of them is more likely to prosper in the neighbourhood of towns: commuting to jobs in towns, pluriactivity of family farms, direct distribution channels to consumers, co-operation among producers, and new development blocs. If and when the most important survival factor of rural areas is interaction with towns, the remote periphery has to make the biggest efforts to maintain communication and transportation facilities. The new infrastructures connect first national and province capitals and university towns. If they are to stretch out to small centres and even more remote areas, strong political pressures and special measures will be needed.

The third policy area influencing rural development has been the rural welfare policy, which was the by-product of general welfare reforms. As has been stated before, it had a strong balancing impact on regional development in the 1980s. It created public sector jobs in rural centres and provided services that were needed by the population. The mutually enforcing influence of both the welfare state cuts and the adaptation to European markets decreases the probability of frog leaps within remote rural areas. Rather than brisk frogs eager to leap, the local communities are more likely to freeze or hibernate in wait for warmer times.

What is needed, is a policy programme consisting of elements that answer the problems of these peripheries and keep the frogs from freezing. In addition to supporting some small scale farming and maintaining the landscape, which is the recipe of European Union common agricultural policy, some service structures must be retained. This probably is a kind of rural welfare policy as a part of the European Union cohesion policies and its social and structural programmes. However these could be combined with creating networking infrastructure. In the peripheries, maintenance of isolated administrative or policy-sectors would be expensive and also a waste of resources, when the same structures could be serving several aims. The conversion of Finland's regional policies into European Union programmes has been going on successfully, according to many evaluations. However, this task is administrative and technical in nature. To create a coherent policy for remote peripheral rural areas is another task, and to gather up political will and power to push those policies through the EU apparatus, is yet another task. For this the rural blocs have to find strength in each other and to make some new friends. They have to find allies in town and they have to find their kind in other countries, and to start building international co-operation. This cannot be accomplished by a lone village telecottage with a handful of enthusiastic activists. There is, however, some evidence of results that are brought about by the skillful and persistent building of networks. For example in Finland's south-western archipelago a network of home-based flexible work has been constructed with co-operative efforts of the university, labour district, local authorities, telecottages, individual producers, and the unemployed persons etc.

In the beginning of this chapter I suggested that the core-periphery hierarchies have not melted into the air, but changed their nature. Being in a periphery has always something to do with economy, politics, and culture, to use Hechter's (1975) terminology. As new social orders are established, however, the key dimension of peripherality may change, "key" here meaning something you need to open the door. Successful political action will focus on that key. In the peasant society the key was the right to land. In the Welfare Society it was the participation in reforms that were linked to economic growth. In the emerging order the key dimension seems to be linkages and connections. If you are connected, wherever you are geographically, you are participating in the networks of the cores. If you are disconnected, you are in periphery, even if you are physically located at the heart of London.

References

Dodgshon, R. A. (1987), *The European Past. Social Evolution and Spatial Order*, Macmillan Education Ltd, London.
Commission of the European Communities (1988), *The Future of Rural Society*, Brussels.
Cronberg, T., Kolehmainen, E. and Lehikoinen, A. (1990), *Tietotuvat Suomessa*, (*The Telecottages in Finland*, Ratko report.) Ratkon julkaisuja nro 4, Ruvaslahti/Joensuu.
Galtung, J. (1966), 'Rank and Social Integration. A Multidimensional Approach', In Berger, Zelditch and Anderson (eds), *Sociological Theories in Progress*, Vol.1. Boston, 145–198.
Galtung, J. (1968), 'A Structuralist Theory of Integration', *Journal of Peace Research*, 1968/4, 375–395.
Koskinen, T. (1985), 'Finland – A Forest Sector Society? Sociological Approaches, Conclusions and Challenges', In Lilja, K., Räsänen, K. and Tainio, R. (eds), *Problems in the Redescription of Business Enterprises*, Helsinki School of Economics, Studies B-73. Helsinki, 45–52.
Malinen, P., Häkkilä, M. and Jussila, H. (1993), *Finland's national rural policy facing the challenge of European integration*, Research Institute of Northern Finland, University of Oulu, Oulu.
Marsden, T. Murdoch, J., Lowe, P., Munton, R., Flynn, A. (1993), *Constructing the Countryside*, UCL Press, London.
OECD (1993), *What Future for Our Countryside? A Rural Development Policy*, OECD, Paris.
Oksa, J. (1993), 'Uuden ja vanhan viestinnän rajalla: Kontiolahden viestintäprojekti maaseudun tutkijan silmin', (On the Borderline of new and old Communication.) Kontturi, P., *Maaseudun tiedonvälityksen kehittämisprojekti. Loppuraportti*, (Append-

age to Final Report of the Kontiolahti Rural Communications Development Project), Kontiolahti 20.1.1993.
Oksa J. (1994), 'Villages Coping on the North-eastern Margin', In Symes, D. and Jansen, A. (eds), *Agricultural Restructuring and Rural Change in Europe*, Agricultural University, Wageningen, 270-281.
Pyy, I. (1994), 'Paikallisen politiikan haave ja arki', (Vision and reality of local politics), In Oksa, J. (toim.), *Syrjäisen maaseudun uudet kerrostumat*, (New Activities in Remote Rural Areas), Joensuun yliopisto, Karjalan tutkimuslaitoksen julkaisuja, University of Joensuu, Publications of Karelian Institute, N:o 110, Joensuu, 38-49.
Qvortrup, L. (1990), Community Teleservice Centres in North-Western Europe: A February 1990 Update, *Newsletter on Community Teleservice Centres*, April 1990, International Association of Community Teleservice Centres, 24-33.
Rural Policy Committee (1992), *Rural Policy in Finland*, Ministry of the interior & Ministry of agriculture and forestry, Helsinki.
Sahlins, M, and Service, E.A. (1960), *Evolution and Culture*, U Michigan Press, Ann Arbor.
Sireni, M. (1994), 'Pluriactivity of farming households in Eastern Finland. A Case Study of two Rural Municipalities in North Karelia', *NordREFO* (Nordisk Institut för Regionalpolitisk Forskning), forthcoming.

CHAPTER 11

Internationalization from the European Fringe: the Experience of SMEs

Heikki Eskelinen
University of Joensuu

Leif Lindmark
Umeå University

11.1 The Role of SMEs in the Nordic Countries

In the Nordic countries, as in many other countries, economies of scale was the leading economic principle in the 1960s. In Sweden and Finland in particular it was the decade of large firms. Then came the 1970s with economic crises. Firms in several industrial sectors were badly hit by changes in the economic climate. Mainly it was large rather than small firms that had problems. A result of the crises faced by large firms was a growing political interest in small and medium-sized enterprises (SMEs) and entrepreneurship. The formation of new firms increased as did the number of people working in SMEs. Thus the 1980s can be labelled the decade of SMEs.

For the period 1985-89 the role of Swedish SMEs in the job creating process has been studied in detail (Davidsson, Lindmark & Olofsson 1993). The study clearly shows that the role of SMEs in the job creating process is impressive: seven out of ten net new jobs were created by SMEs. In the study SMEs were defined as autonomous firms with less than 200 employees.

The downturn in the world economy during the first years of the 1990s also affected SMEs in the Nordic countries, but statistics suggest that they were hit less hard than large firms. As a result the share of people working in SMEs increased. There are also clear indications that the positive trend for SMEs will continue during the rest of the 1990s.

The positive development of SMEs can partly be explained by the increasing specialization and division of labor in industry. Other factors precipitating a positive development of the SME sector are changes in technology and the growth of the service sector. The positive progress is also a result of SMEs having been successful in adapting to changes in the economic environment including increased international competition. More and more SMEs are becoming internationalized.

The Nordic economies are small, open economies. In spite of their location in the periphery of Europe the foreign trade in relation to GDP is higher than that for larger countries located in more central parts of Europe. The Nordic countries, therefore, are very dependent on economic and institutional changes outside their own borders. Changes in Eastern Europe have severely hurt the Finnish economy, and the creation of the internal market inside the EU has resulted in adaptation processes in all the Nordic countries, among which only Denmark has, for a longer time, been an EU member. Finland and Sweden joined from January 1, 1995, whereas Norway continues to arrange her relations with the EU through the EEA agreement, which has been in force from the beginning of the year 1994. In addition to these institutional changes, other factors precipitating the internationalization of the Nordic economies are e.g. better language skills, increased knowledge about other business cultures and better communications.

In spite of an increasing internationalization of the Nordic economies, it seems that the local/regional environment of the firm has not become less important, but is rather of increasing importance for the internationalization of SMEs and for their competitiveness. This may sound paradoxical at a time when globalization and improvements in transport and communications have provided firms with more possibilities than ever before for reorganizing and shifting their operations to different locations.

Interdependencies between the dynamics of a firm and its commitment to a certain local or regional environment have been the subject of a great deal of research in recent years. The emphasis given to the role of the spatial environment can be derived from ongoing changes in competitive conditions and strategies of firms. Intensified competition leads to specialization, which increases demand for external resources of firms, such as subcontracting and R&D competence. Successful regions and cities develop into specialized operational environments which can provide firms with strategic resources, and consequently, a competitive advantage. External resource utilization seems to be of great importance, especially for SMEs. As a result, the fates of firms and regions become intertwined in an even more intimate way than before.

The rationales for the accentuated role of a local environment and the dynamic interplay between the development of firms and regions have been analyzed by utilizing a variety of theoretical concepts and frameworks. They focus on institutions, industrial systems and transaction costs, or technological changes and learning (see Storper 1995). Industrial districts and milieus have been used as key-words to describe these environments. The former refers to specialized, usually non-metropolitan regions where (small) firms benefit from economies of scale and scope through cooperation (see e.g. Brusco 1992). This cooperation is argued to be based not only on pure economic incentives, but also on the distinctive social and cultural traditions of each region. According to the milieu approach, a firm is said to benefit from its embeddedness in the local business environment in basically two ways. A favourable milieu can diminish uncertainty as well as support a collective learning process contributing to innovations (see e.g. Camagni 1991).

A local or regional business environment is, by its very nature, a dynamic concept. Therefore its relevant qualities are not easy to identify although certain infrastructural properties (e.g. proximity to an international airport and university) tend to grow in importance as locational factors in a modern economy (see e.g. Johansson 1993). In particular, the competitive advantages of a spatial environment can be based on tacit knowledge which is created and learnt through interrelationships between economic actors. The strategic value of this kind of knowledge stems from the fact that it is only to a limited extent imitable and substitutable. It has been intertwined with local "untradeable interdependencies" which can furnish the firms with an absolute advantage in their own specialized field of activity (see Storper 1995).

In empirical research, a spatial environment is used to refer to different geographical and institutional contexts. Industrial districts, for instance, have been conceptualized both as daily urban regions and as larger functional units. In addition, a relevant operational environment can be interpreted in terms of countries and national economies, as Porter (1990) has done in his analysis of the competitive advantage of nations.

This chapter examines the significance of the local environment for the internationalization of SMEs in regard to three Nordic countries, Finland, Sweden and Norway. This research task has received its impetus from the expectations that the spatio-economic characteristics of these countries might weaken their possibilities of adapting themselves to the economic competition, which will probably intensify as a result of the ongoing internationalization and integration processes.

The policy context of the study is discussed in next section, and in section 11.3, the empirical setting of the comparative study is briefly presented. In section 11.4 data about international activities of Nordic SMEs are presented showing that a large number of SMEs act on international markets, and in section 11.5 a theoretical framework on the internationalization of SMEs is outlined. Section 11.6 summarizes empirical observations from the Nordic study and relates the findings to the theoretical framework. Finally, some conclusions are derived. They focus on the role of specialized industrial districts in the Nordic context and on the changing spatio-economic characteristics of Nordic peripheries.

11.2 Nordic Peripheries

The Nordic countries (excluding Denmark) are geographically peripheral in the Western European context. The bulk of their economic activities and population is in their southernmost regions and consequently, large areas are very sparsely populated. The economies of these countries, especially in peripheral regions, have traditionally been based on the exploitation and processing of natural resources (timber, fish, ore, and, currently in Norway, oil and gas). These resource-based activities have gradually developed into major production complexes or national industrial clusters, which are still responsible for a major share of Nordic exports (see e.g. Lundberg 1992). Firms belonging to the leading clusters have been active in developing technology and international activities. In fact, they have had a decisive role in the process through which the Nordic economies have acquired their positions among the developed industrial countries.[1] This relative prosperity has, for its part, created the resources for the construction of the Nordic welfare state (see e.g. Commission of the European Communities 1993).

The natural resources which have formed the basis for the traditional specialization of the Nordic economies are spread relatively evenly across vast areas. The dispersed settlement structure derives from this. Although spatial concentration has proceeded in recent decades, the

---

[1]For industrial clusters in these Nordic countries: see Sölvell et al. 1991 (Sweden), Hammervoll & Lensberg 1992 (Norway) and Hernesniemi et al. 1995 (Finland).

transfer mechanisms of the welfare state have also contributed to the consolidation of the existing pattern. These measures have concerned physical and social infrastructure in so wide a meaning, that citizens have been given (more or less) equal opportunities in the provision of services and in income transfers irrespective of where they live. Overall, the Nordic periphery has quite clearly been more prosperous, at least up until the 1990s, than other geographical peripheries of Western Europe.

Current changes in the international division of labor highlight still further the competitive pressures facing developed industrial economies such as the Nordic countries. To an increasing degree, they should be able to orient themselves towards competition by means of new specialized products and skill-intensive technologies. For the sake of continuity alone, national industrial clusters also have major roles in this development, but their strategic competitive advantages keep changing. Man-made production factors grow in importance whereas natural endowments remain more and more in the background. For instance, this trend can be seen in the fact that an increasing proportion of Nordic exports consists of so-called intra-industry trade (see Kajaste 1992). Also, SMEs have to adapt to increasing competitive pressures.

In addition to the above-mentioned structural pressures, the Nordic countries, in particular Finland and Sweden, have been undergoing a deep economic recession in the 1990s. The collapse has been most serious in Finland, whose GNP has diminished by more than one-tenth in three years. This has led to a situation in which the GNP per capita in the more peripheral eastern and northern parts of the country is nowadays below the level set by the criteria used in delineating the objective 1 regions in the European Union. In the debate concerning the current crisis, the very foundations of the Nordic model have been questioned by pointing to profound changes in competitive conditions and institutional structures of the international economy (see e.g. Lindbeck et al. 1993). Prospects of this kind are usually discussed in the context of the EU integration which is expected to accelerate ongoing changes.

Quite irrespective of whether the Nordic model is bound to collapse or not, problems of economic adjustment can be argued to be most severe in the peripheries of these countries. This is not only, nor even primarily due to their geographical remoteness and dependence on resource-based activities and net income transfers (see Eskelinen 1993).

Assuming the increasing role of a local or regional environment in economic dynamics, an even more crucial problem can be assumed to be caused by the fact that functional regions in the Nordic peripheries are small in terms of population and economic activities. Thus their resources are scarce. For instance, labor markets are nondiversified and sophisticated business services are lacking. This raises questions about their competitiveness in an ever more internationalized economy: what kinds of competitive advantages could be developed in these dispersed towns and settlements so that they might provide specialized and internationalized enterprises with a seedbed of success?

11.3 A Nordic Comparative Study

Whatever the concrete forms of interdependency between a firm and its spatial environment might be in various circumstances, internationalization is an indisputable feature of current economic development. It is also, particularly in the form of export activity, a vital strategic alternative to a growing number of small firms (see e.g. Christensen 1991). The relevance of internationalization as a research theme is increased by the fact that it obviously presupposes that a firm possesses special skills and competence. The supply of these resources can be expected to vary according to the spatio-economic characteristics of regions and localities.

Thus, the problems and challenges of these regional economies can be explored by using local SMEs as indicators of current structures and processes. The most important reason for this is the fact that a typical SME has only one locational site and its internal resources are scarce. Consequently, its competitive edge can be assumed to be especially dependent on the external resources available in its spatial environment.

In the study presented in this chapter an SME is defined as an autonomous, locally-owned industrial firm with 10 to 199 employees. Branch plants and subsidiaries are not included because they have an intra-organizational access to non-local resources. The study is limited to the wood-products and engineering industries which occupy key positions in the production structure and exports of the three Nordic countries concerned.

The data are based on detailed interviews with owners/entrepreneurs in 274 SMEs in the two industries.[2] A local environment is delineated in terms of labor market districts. Its role is assessed by comparing SMEs in four different types of regions in the three countries.

The location of industries and the rates of industrialization differ considerably between regions in Finland, Sweden and Norway. This influenced the selection of the target regions. Four different types of regions (large city, industrial town, small town and rural periphery) represent different levels in the hierarchical settlement system, thus illustrating the internal spatio-economic differentiation of the countries. They also differ in terms of accessibility to the main domestic and international markets. With regard to the competitive conditions of SMEs, the most important differences concern distances and transport costs, population density, size of local markets as well as supplies of services and subcontracting. In general, the regions in question can be assumed to provide very different local preconditions for the internationalization of SMEs.

The data are based on a random sample of SMEs in large cities, and in other regions on a census. They were collected by interviewing owners/entrepreneurs in the spring and summer of 1992. The interviews, of one to three hours duration, were based on a common questionnaire. For purposes of comparability, the available choices of replies were standardized. Most firms contacted for the study were willing to cooperate; less than 20 firms refused to participate in the interviews. Of the 274 SMEs, 97 are in the wood-products and 177 in the engineering. There are 80 Finnish, 101 Swedish and 93 Norwegian firms. Some basic information about the SMEs is presented in Table 11.1.

The SMEs in the study employed about 9000 persons in 1991. The median size was 21. Their total turnover was almost one billion US dollars. The criteria used in the selection of the SMEs and the size of the sample (number of firms) provide grounds to claim that observations

---

[2]For the framework of this comparative Nordic study on the internationalization of SMEs: see Christensen et al. 1991; Christensen & Lindmark 1993. On the empirical findings: Lindmark et al. 1994; Eskelinen et al. 1994; Eskelinen & Vatne 1995.

based on these firms are representative of the SME sector in the wood-products and engineering industries in the Nordic countries.[3]

Table 11.1. Basic information about SMEs studied

	Finland	Norway	Sweden	Total
Number of firms	80	93	101	274
- wood-products	23	39	35	97
- engineering	57	54	66	177
Employment (median)	19	21	24	21

11.4 Export Sales

Finland, Norway and Sweden countries export 30–40 per cent of their respective GDPs. Even though large firms dominate export sales there is a large and increasing number of SMEs that act on international markets. In e.g. Sweden there were 33 000 exporting firms in 1991, of which 28 000 were SMEs. The exports of the majority of the SMEs are small, but because of the large number of SMEs, their total export value is rather impressive. Autonomous SMEs account for just under 20 per cent. Indirect exports are not included in these figures (Lindmark, Lindbergh and Palm 1995).

Products and services are exported by SMEs of all sizes. More than 20 per cent of the export value from autonomous SMEs is delivered from firms with less than five employees. That is about the same share as firms with 100–199 employees. There are variations between industries. In the wholesale and retail trade industries, a relatively large proportion of exports comes from smaller SMEs. In the manufacturing

---

[3]Most of the data is cross-sectional and confined to the SMEs which existed in 1992. This implies that those SMEs which have been taken over by larger firms or which have gone out of business have been excluded although these changes could also be indicators of the significance of a local environment for competitiveness.

sector, on the other hand, the larger small firms are the more important exporters (Lindmark, Lindbergh and Palm 1995). Anyhow, the data clearly show that Swedish SMEs are established in international markets and that it is not only the larger SMEs that export. Export barriers can also be overcome by very small SMEs.

The internationalization of SMEs is usually limited to exporting and operations directly supporting it (e.g. the establishment of marketing subsidiaries).[4] This also applies to the Nordic SMEs under consideration here. For instance, only a few of them have internationalized by making direct foreign investments.

Of the SMEs studied almost six out of ten were exporters. The share of exporting firms was somewhat higher in Norway and Sweden than in Finland, whereas the share of exports of the total turnover was in Norway much lower than in the two other countries concerned. Measured as a percentage of the sales of the individual firm the average share of exports is between 16 and 22 per cent. Taken together the figures in Table 11.2 below suggest that Swedish SMEs are more export-oriented than SMEs in the other two countries.

Table 11.2. SME's exports by country

Country	Share of exporters	Per cent of total sales of SMEs	Average exports of each SME
Finland	45	33	17
Norway	63	22	16
Sweden	60	35	22

The data also show that it is a rule rather than an exception that exporting SMEs sell to more than one country. More than one fourth of the exporters had customers in five or more countries. This was especially the case for Swedish SMEs. In Table 11.3 the five most important export countries for each SME in the study are presented.

---

[4]In contrast, the analysis of the internationalization of larger firms usually concentrates on forms of operations which go beyond traditional commodity trade.

By and large the Nordic SMEs have the same pattern of exports as large firms. Export sales are very much concentrated to Western Europe. However, exports to neighbouring Nordic countries seem to be of greater importance for SMEs than for large firms. The Swedish market is of especially great importance for Norwegian SMEs. SMEs do export also to other parts of the world. For many niche-oriented and high-technology firms the most interesting markets are the US and/or Japan.

Table 11.3. SME's exports to different regions based on information about their five most important export markets (number of firms).

Country / Market	Finland	Norway	Sweden	Total
Norway	15	-	32	47
Finland	-	8	25	33
Denmark	6	12	15	33
Sweden	23	45	-	68
Germany	13	12	28	53
UK	11	14	21	46
Rest of Western Europe	18	23	49	90
Eastern Europe	2	3	3	8
USA	6	4	6	16
Japan	3	1	2	5
Rest of the world	12	10	12	34
Number of exporting SMEs	36	58	61	155
Number of SMEs	80	93	101	274

Source: Lindmark et al. 1994

11.5 Internationalization of SMEs

Small enterprises often suffer from a lack of internal resources, managerial as well as financial. However, the correlation between size and the level of globalization is not clear-cut. In a study of the internationalization of young, technology-based Swedish firms only a weak correlation between size and internationalization was found, which indicates that factors other than firm size are also of importance (Lindqvist 1991). One such factor is the entrepreneur of the firm, his (or her) ability, experience and network. The importance of the entrepreneur for the globalization of SMEs cannot be underestimated.[5]

Other impeding factors to the internationalization of Nordic SMEs are geographical distance and transport costs to foreign markets. Finland, Norway and Sweden are located in the periphery of the European market. Cultural differences, lack of market knowledge including problems identifying partners and customers abroad, and strong foreign competition are also impeding factors. Foreign competition was a major impediment to Finnish and Swedish SMEs in starting or increasing export sales, especially in the late eighties and the beginning of the nineties when the value of the Finnish mark and the Swedish krona were much higher than today.

A number of internal and external factors precipitate and sustain the internationalization of SMEs. Of the internal factors the specialization and niche-orientation of the firms seem to be of special importance. These factors are very much related to another factor, limited home market potential. The more niche-oriented the firm is, the stronger is the need to go international. Other internal factors precipitating and sustaining internationalization of SMEs are language skills, the prospects of taking advantage of economies of scale and strategic risk-diversification.

The market situation at home and abroad can be either a pull or push factor for internationalization. Decreasing demand on the home market and/or strong competition at home can force SMEs to start and develop international activities. There is no doubt that the strong downturn in the economies of the three Nordic countries in the beginning of the

---

[5]On the analysis of the internationalization of SMEs: see e.g., Cappellin 1990, Christensen & Lindmark 1993 and Vatne 1995.

nineties has been such a push factor. On the other hand a large market potential and limited competition abroad is an important pull factor creating conditions for SMEs in Finland, Norway and Swden to go international – especially those in the high-technology and niche-oriented sectors. Another important factor is access to advanced customers. Being in the same markets as its main competitors gives the firm better possibilities to follow and adapt to changes in demand and product development.

The internationalization of a firm is most commonly described and analyzed as a gradual learning process, in which the firm gathers information and experience from international operations and thus strengthens its commitment to them. Internationalization is usually conceived of as a rational process where economic, managerial and organizational resources are engaged step by step in increasingly committed international activities. This framework, a stage model, is not a theory in any strict sense. In particular, it does not include any explanation of how a firm advances from one stage to another.

The stage model has been criticized both on theoretical and empirical grounds. For instance, it has been emphasized that operations required by particular activities do not follow each other in a certain sequence of events, but rather are chosen on the basis of situational strategic considerations. Empirical studies of SMEs have shown that many small firms start and develop export activities without preplanning. In addition, the small size of the enterprise makes it more responsive to random and unique factors. For instance, its first export deliveries are often based on unsolicited orders. The SME initiation of international activities can therefore often be seen as an entrepreneurial process in regard to the way that international contacts are established and external resources are mobilized.

Studies of SMEs also suggest that a number of SMEs become internationally engaged in a more rapid and direct way than anticipated by the stage model. The time lag between the birth of the enterprise and the initiation of international activities has been narrowed. Some firms, especially high-tech and niche-oriented firms, start to export more or less directly after the formation of the firm. The national market is already too small for these firms to survive in the start-up phase. In these cases we can talk about an entrepreneurial process in three dimensions at more or less the same time: the start-up process, the product development process and finally the internationalization process.

Furthermore, it has been noted that the stage model has been developed primarily on the basis of experiences of large corporations which have been successful in their international operations. Therefore it emphasizes the role of the internal resources of the firm, something which undermines its relevance in analyzing the internationalization of SMEs.

According to Miesenbock (1988), a conclusive theory of small business internationalization is far from being available. In an extensive literature review Miesenbock discusses a number of shortcomings in the existing theories and points out that two important factors are not discussed at all.

Firstly, the literature does not take into consideration the large number of small firms positioned in industrial markets and the special business conditions that characterize these markets. The often very tight linkages that make up the international value added chains imply high barriers to entry for firms looking for new business opportunities. However, empirical studies of Swedish SMEs suggest that one precipitating and sustaining factor for internationalization is the role as a subcontractor to MNEs. There are also indications that the internationalization of SMEs is precipitated and sustained by the internationalization of customers.

Secondly, the theories almost exclusively focus on the internal resources of the firm available to sustain the internationalization process. The use of external resources – private and public – to overcome barriers are with few exceptions completely left out. Considering the discussion about the importance of external resource utilization in the small firm literature this is remarkable.

The scant internal resources of SMEs limit their market power and orientation towards foreign markets. Lacking the specialized skills it needs, a small enterprise has to find external resources to support its internationalization process. This can be assumed to accentuate the significance of strategic behaviour and the local environment.

Thus, gradual learning and commitment through experience, strategic behavior and dependence on the characteristics of a local environment as well as unique, situational influences can be linked in different combinations in the development of international operations. In general, we can assume that internationalization demands strong internal resources and/or an active search for external resources from an SME. The interdependencies between these different factors are outlined in Figure 11.1.

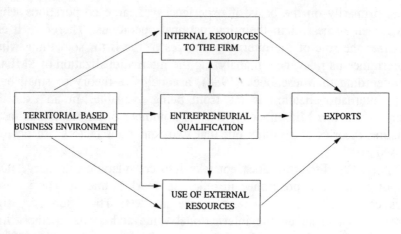

Figure 11.1. A framework for analyzing the internationalization of an SME

11.6 Model and Reality

The framework in Figure 11.1 has been influenced by models of industrial districts. It is assessed in the present study on the basis of data concerning three countries, and it therefore is possible to use a national environment as a point of comparison and reference to a local environment. Another reason for this was mentioned earlier: the industrial development of the Nordic countries has to a considerable degree proceeded in the context of national clusters.

11.6.1 Market Orientation

The level of internationalization varies considerably between exporting SMEs. In the Nordic study there are firms that are large exporters, while others export only a minor share of their turnover. The export strategy also differs among SMEs. Some base their export sales on a penetrating strategy, others use a skimming strategy. The former implies extensive exports to one or a few countries, the latter exports to many markets, but with only minor export sales to each foreign market. These and other differences between exporting SMEs make comparisons between exporters and non-exporters problematic. There are obvious risks that differences between the two groups will not be observed if firms with major export activities are not separated from those with only minor export sales.

Thus, in the following the group of exporters has been divided into two groups, active and experimental exporters. The division is based on export sales and type of export strategy. SMEs exporting 20 per cent or more of their turnover and/or exporting to at least three countries have been classified as active exporters. Other SMEs involved in exports have been classified as experimental exporters. Of the SMEs in the Nordic study 40 per cent were classified as active exporters, 17 per cent as experimental exporters and 43 per cent as home-market oriented firms. The latter had no exports at all.

In Table 11.4 some data based on the three categories are presented. Of the three countries in the study Sweden has the largest share of active exporters, which probably can be explained by a stronger tradition among SMEs to act internationally than in Finland or Norway. Finland has fewer exporting SMEs in comparison with Norway. Yet those Finnish firms which have gained access to foreign markets export a greater proportion of their production. This structural pattern is in line with the view that export thresholds are higher for Finnish firms due to cultural and geographical reasons. Consequently, when a firm surpasses these thresholds, it is more committed to exporting.

The structural features of the locational environment of an SME are reflected in the regional distribution of its sales. The basic setting is clear-cut: given small local markets, production has to be sold in other domestic markets or exported. In the present comparison this contributes to the fact that the share of the exporting SMEs is higher in rural peripheries than in other regions. Another factor supporting this somewhat surprising observation is the sectoral structure of industry. Resource-based wood-products industries such as sawmills traditionally play a major role in the remote peripheries of Finland, Sweden and Norway. These industries are by tradition strongly export oriented.

The adjustment of the SMEs to the local environment is also discernible in the fact that a relatively higher percentage of the firms in the rural peripheries produce their own products while a lower number of them have found a market niche in subcontracting. Industrial towns represent a contrasting pattern in the sense that the SMEs there have relatively lower export shares whereas they have aligned themselves to act as subcontractors to local industries.

Table 11.4. Active exporters, experimental exporters and home-market oriented SMEs by country, industry and region (%)

	Active exporters	Experimental exporters	Home-market oriented SMEs	Total
Finland, n=80	31	14	55	100
Norway, n=93	37	26	38	100
Sweden, n=101	50	11	40	100
Wood-products, n=97	34	20	46	100
Engineering, n=177	43	15	42	100
Large city, n=71	39	17	44	100
Industrial town, n=72	38	18	44	100
Small town, n=69	33	15	52	100
Rural periphery, n=62	50	18	32	100
Exporters, n=155	70	30	-	100
Total, n=274	40	17	43	100

Source: Lindmark et al. 1994

11.6.2 Internal Resources and Export Orientation

The models of internationalization are very much focused on internal resources of the firm and the commitment of internal resources to the internationalization process. Data from the Nordic SME study support the notion of the importance of internal resources to internationalization and international competitiveness.

First of all, active SME exporters are larger than home-market oriented SMEs. They are also more capital intensive, have a higher technological level and have automated their production system more (Table 11.5 below). These observations suggest that successful international activities are not only a question of market activities, but also of production and production facilities. The more advanced the SME, the better the conditions for exports and international competitiveness.

Table 11.5. Internal resources to the SME by market orientation (%)

	Active exporters	Experimental exporters	Home-market oriented SMEs	** .01 * .05
Employment				
- median	30	29	16	**
- arithmetic mean	45	39	24	**
Turnover, million FIM				
- median	17	10	6	**
- arithmetic mean	24	16	11	**
Division of labor (scale 1-5)				
- number of departments	4,6	4,1	3,2	**
Production process (scale 1-5)				
- technological level	2,9	2,7	2,3	**
- automatization	2,6	2,2	1,7	**
Owner/Entrepreneur/experience (%)				
- large company	55	49	47	
- consultancy business	19	20	13	
- foreign company	32	29	16	*
- other industry	66	47	47	**
Owner/Entrepreneur/higher education (%)				
- at least 4 years	23	21	9	
- 2-3 years	36	41	36	
- less	40	39	56	
Owner/Entrepreneur/language skill (%)				
- English	87	84	66	**
- German	60	44	29	**
Personnel				
- share of white collar (%)	29	22	20	**
- change in level of education (%)				
+ strong increase	14	9	3	
+ minor increase	47	43	35	
+ constant or decrease	39	48	61	**
- education (% of white collar)				
+ university	14	12	13	
+ other higher education	37	24	30	
- experience (% of white collar)				
+ large company	20	20	21	
+ consultancy business	6	3	4	
+ foreign company	10	6	9	
- language skill (% of white collar)				
+ English	58	46	49	
+ German	22	12	16	

Cross-tabulation χ^2; one way analysis of variance, F-test. Internal division of labor and the design of the production process have been measured on a scale from 1 to 5 where higher values indicate a more comprehensive division of labor in the firm's functions and a more advanced production process respectively.

The internal human and economic resources of an SME are, of course, dependent on its size and performance. The larger the firm, the more internal resources. However, it can be argued that the qualities of these resources (e.g. the competence and skills of the labor force) and their compatibility with the strategy of the firm are even more important. Usually the entrepreneur himself or herself is the key human resource of an SME.

This theory is supported by empirical data from the Nordic study. There are significant differences in education, and experiences between owners/entrepreneurs in active exporters and those in home-market oriented SMEs. The owners/entrepreneurs in active exporters have on average a higher education, better language skills and more work experience from other firms, e.g. from companies abroad. There are also differences between the white collar workers in the two types of SMEs, but these differences are small compared with the differences between the owners/entrepreneurs. However, active exporters also seem to have more resources than home-market oriented firms when it comes to white collar workers.

11.6.3 External Resources and Export Orientation

An analysis of the use of external resources shows that the similarities between the three categories of SMEs – active exporters, experimental exporters and home-market oriented firms – are larger than the differences (Table 11.6). These observations could be interpreted as evidence that local resources are of no importance for the international activities of SMEs.

For several reasons such a conclusion may be wrong. First of all, the quality and the frequency of the contacts are only partly included in the analyses. Secondly, there are alternative interpretations of the similarities in the use of external resources. The observation that owners/entrepreneurs in active exporters do not have more contacts with subcontractors, customers and competitors than their counterparts in home-market oriented SMEs can probably be explained by differences in size between the firms. With increasing size people in the firm other than the owner/entrepreneur can also have these contacts. This conclusion is supported by the fact that the divison of labor in the firms increases with size.

Nor can differences be found in the share of firms using regional business services. While home-market oriented SMEs mainly use regional services, active exporters also buy these services at the national and international level. The more frequent use of national and international resources is also supported by the fact that among active exporters the most important partner for cooperation is located outside the region more often than for other SMEs. Of active exporters, experimental exporters and home-market oriented SMEs, 30, 47, 54 per cent respectively report that their most important partner for cooperation is located in the same region. These figures can be compared with the fact that almost one fourth of the active exporters have their most important partner located in another country. The same pattern emerges in an analysis of the location of subcontractors to the firms.

Another empirical observation is that it is more common among owners/entrepreneurs in active exporters to have social contacts with people in the region and contacts with people in other organizations than suppliers, customers and competitors. Many of the contacts of importance for SMEs in their international activities can be assumed to be of another kind than market relations.

It is worth noting that a higher share of experimental than active exporters buy business services in the same region. This could be interpreted as support for the thesis that the local environment is more important for SMEs in the first phases of the internationalization process than in later phases. With increasing export activities the center of gravity of the network of the firms is moved from the regional to the national or international level. This movement is a result of SMEs adding new contacts to already existing ones, not a consequence of SMEs substituting regional contacts for national and international ones.

Table 11.6. Use of external resources by type of SME

	Active exporters	Experimental exporters	Home-market oriented SMEs	**.01 *.05
Contacts by entrepreneurs				
at least every week in % of firms				
- professional contacts in the region	55	60	55	
- social contacts in the region	46	39	41	
at least once a month in % of firms				
- subcontractors	57	78	57	*
- competitors	32	38	31	
- customers	80	82	70	
- other organizations	41	35	29	
Forms of cooperation, in % of firms				
- joint-owned firms	31	30	25	
- long-term cooperation based on contracts	55	52	47	
- other long-term cooperation	58	61	65	
- temporary cooperation	75	80	76	
Relations in cooperation				
- number (aritmetic mean)	2,2	2,3	2,0	
- localization of most important partner				
+ local environment, in % of firms	30	47	54	**
+ abroad, in % of firms	24	3	2	
Subcontracting/turnover, %				
- subcontracting from firms in the region	9	13	8	
- subcontracting from others	13	6	8	*
- subcontracting to firms in the region	7	12	36	**
- subcontracting to others	39	33	29	
Business services in the local environment in % of firms buying a certain service	84	85	90	
- Accounting and book-keeping	75	80	76	
- ADP	65	70	70	
- Advertising and marketing	76	76	79	
- Transportation	49	45	44	
- Management consulting	55	70	61	
- Technology consulting	72	76	76	
- Legal consulting	53	68	68	
- Staff training	84	89	94	
- Financial services				

Note: see table 11.5.

11.6.4 EU Exporters and Other Exporters

For many SMEs in the Nordic countries the internal market of the EU is of great importance. With the ongoing internationalization process, supported by the integration agreements between these countries and the EU, the importance of the internal market will probably be of even greater importance for Nordic SMEs in the future. To analyze whether there are differences between SMEs exporting to the EU and other SMEs the firms in the study were grouped according to their market orientation: EU-exporters, other exporters and non-exporters.[6]

As already emphasized, the export activities of an SME presuppose the development of the necessary resources. On the other hand, the resources of an exporting SME are upgraded as a result of competition in export markets. Thus, strategic internationalization and the effects of competition are intertwined with each other. Their joint repercussions can be seen in the resources of the SMEs.

A Nordic SME exporting to the European Union is on average larger than its counterparts and has more internal resources. It emphasizes R&D activities but it does not regard price as the primary measure of competition. Yet it is not more profitable than other SMEs. Also the EU-exporter is more likely to have cooperative partners outside the local region and outside the country.[7] Yet it is basically similar to any other SME since it does not purchase more subcontracting or producer services from other local firms, nor does it cooperate with them more often.

As far as the threats of competition facing the peripheries are concerned, the findings suggest that the Nordic SMEs exporting to Western European markets have already to a great extent adjusted themselves to the requirements of the internal market. The effects of integration, which were analyzed on the basis of the assessments of the entrepreneurs, seem to have more influence on those firms without exports to

---

[6]The criterion of an EU-exporter is that at least one of its five most important export countries is an EU member.

[7]Hanberger (1993) has observed that exporting Swedish SMEs cooperate somewhat more often with other local firms, especially in subcontracting. The present data set does not support this.

the EU. In particular, the greatest repercussions are expected by those SMEs which only have domestic sales. In general, the anticipations of the entrepreneurs in the rural peripheries are not on the average more negative than those of the entrepreneurs in the other regions.

11.6.5 Localization and Internal and External Resources

The analyses so far support the idea that the internal resources of the firm, and especially the resources of the owner/entreprepreneur, are of great importance for the internationalization of SMEs. The importance of local, external resources for internationalization cannot find the same support in the empirical data. The question is therefore to what extent differences in resources in the local environment affect internal resources of the firm and the use of external resources.

No systematic regional differences were found in the performance of the SMEs in the sense that firms in a certain type of local environment on average were more profitable. The main generalization about the internal resources of the SMEs[8] is much along the same lines: the SMEs in the large cities, industrial towns, small towns and rural peripheries cannot be put into a rank order by region. In particular, firms in smaller and more remote settlements are not on average weaker in terms of their internal resources than firms in more urban locations. The constraints set by a location are seen most discernibly in the fact that there are relatively fewer independent, locally-owned SMEs in remote regions which have also remained generally less industrialized.

Firms have many possibilities for gaining access to external resources. They can buy various services on the open market or acquire them through short-term or long-term cooperation with other firms and organizations. Cooperation can be based on different kinds of institutionalized arrangements or informal links. The latter, for obvious reasons, are very much conditioned by the personal relationships and social role of the entrepreneur concerned. As outlined in Figure 11.1, the resources and local environment of a firm and the problem-solving capacity and contacts of an entrepreneur tend to be intimately intertwined

---

[8]Internal resources refer to the educational level of employees and entrepreneurs, their professional experience and skills as well as to the technology and internal organization of the SMEs concerned.

in the process through which an SME achieves its strategic resources. In this case the potential sources of external resources of the SMEs from a local environment and from more distant locations were surveyed in detail; the forms of and motives for inter-firm cooperation, subcontracting, links with clients, customers and competitors, the use of producer services as well as the social contacts of the entrepreneurs were examined.

Not unexpectably, it was found that the qualities of the local environment of an SME, and the size of the local markets in particular, are clearly reflected in the sources of its external resources. The larger the local economy in which the SME operates, the greater its share of subcontracting and producer services from local suppliers and the greater its share of cooperative relations with local partners. The deviations from this pattern are based on the peculiarities of the provision of services. For instance, with the exception of export consultancy, the use of public services is most common in rural peripheries where these services have been established in the name of regional economic policies.

The most important issue on external resources concerns possible regional differences in their utilization. In this respect, the findings are not well in line with the expectations. In general, the differences are only minor and they do not seem to be linked in a straightforward way to any hierarchical pattern.

It is worth emphasizing here that the broad generalizations above derive from the comparison of the 274 SMEs in different types of local environments on the basis of the combined Nordic data set. The same interview material can be used for comparing the resources of the SMEs in different countries. In general, this comparison strongly suggests that the dissimilarities by country (Finland, Sweden, Norway) are more important than those by type of region (large city, industrial town, small town, rural periphery). These dissimilarities are clearly seen, for instance, in profitability, in some internal resources (e.g. language skills), in subcontracting chains and in forms of inter-firm cooperation. Thus, the national peculiarities of macro-economic conditions and industrial systems seem to condition the resources and performance of the SMEs to a greater extent than the characteristics of a local environment.

11.7 Conclusions

The development of a local or regional business environment and its implications for regional economic growth have been illustrated in many studies based on success stories. This was not the case in the present study, where the research setting was formulated for purposes of a systematic comparison. The aim was to analyze the internationalization of small autonomous local industrial enterprises, and especially the internal and external resources needed for it. Since Finland, Sweden and Norway have relatively similar roles in respect to Western European markets and their internal spacio-economic structures resemble each other, it was possible to construct a quasi-experimental research setting.

11.7.1 Local Environment

Although the regions under consideration were not chosen as supposed illustrations of Nordic industrial districts or favourable local milieus, the factors commonly discussed under these headings influenced the research setting. The study received much of its impetus from the assumption that the qualities of the local environment are relevant to the competitiveness of an SME. If the accessibility, quality and costs of the resources required in the internationalization of SMEs are affected by the characteristics of the community or region concerned, SMEs closer to main markets in larger and more versatile centres tend to strengthen their competitiveness as the process of internationalization proceeds. The analysis of the resources of the SMEs represents an effort to evaluate and qualify this fairly straightforward presupposition.

With regard to this background, the basic message of the empirical findings is fairly clear: the internationalization of the SMEs – for instance, the share of exporting firms and the percentage of exports in the turnover – does not vary according to the qualities of a local environment in any conclusive way. Moreover, the growth and internationalization of a typical Nordic SME does not seem to have been dependent on a local enterprise network or business service infrastructure.⁹ Rather, it could be said that a lack of dynamic specialized industrial districts is one structural characteristic of the current economic crisis in the Nordic countries.

Several reasons have probably contributed to the lack of any direct link between the characteristics of a local environment and the internationalization of SMEs. Basically, the developments in spatial and industrial structures have had to adjust to each other. This is most discernible in the SME sector of the rural peripheries.

Firstly, relatively unfavourable locational conditions such as remoteness from the main markets have resulted in the fact that the most remote regions have remained less industrialized. In addition, their production relies more directly on the exploitation of natural resources. Secondly, most SMEs in the present data set produce relatively simple products, and, consequently, long chains of production with specialized subcontractors are not required. The third salient feature is that many of these SMEs are in those industrial sectors which have long-standing traditions in international operations. The resources of these national industrial clusters have also been of decisive importance in the establishment of the international contacts of many SMEs. Yet their utilization seems to depend to a major degree on the skills and background of the entrepreneur concerned. His or her earlier work experience and personal contacts in several cases form the distinctive competitive advantage of the firm: the "know-how" of internationalization evolves on the basis of "know-who". It is common that general business conditions are discussed in local gatherings of entrepreneurs, but the professional contacts needed to develop a firm are primarily non-local. The most important single reason for this pattern is simply the fact that there is no specialized local competence available in smaller settlements and thinly populated rural areas. Thus a position in the national networks of a relevant industrial sector or cluster has been for many Nordic entrepreneurs more important than a role in a local environment.

This comparative study is based on a cross-sectional data set and includes firms with widely different motives for and processes of internationalization. It is obvious that the above findings do not suffice to

---

⁹In fact, there is only one clear exception to this among the twelve regions concerned. Wood-products industries in the industrial town of Skellefteå in Sweden represent several features of this model: see Lindmark 1993.

reject the hypothesis regarding the increasing significance of a local environment in the dynamics of modern enterprises. Yet they strongly suggest that country- and region-specific peculiarities in industrial systems and institutions should be considered in assessing the models of industrial development. For instance, advantages derived from agglomeration are mostly lacking in Nordic circumstances. In the whole of Finland there are about the same number of industries as in the Emilia-Romagna region in Italy.

In this context, national differences generally refer to a model of organizing and controlling business enterprises which is dominant in a society. These business recipes (see, e.g., Räsänen & Whipp 1992) reflect modes of business behavior which have been successful in certain circumstances, and thus they institutionalize specific economic rationalities. In addition, they are in many cases typical of a dominant sector, the forest sector being a most representative Nordic example of this. The present study suggests that a spatio-economic structure can be considered to be one of the factors conditioning business recipes, or at least certain organizational differences derived from it (e.g., concerning cooperative relations) are comparable with the dissimilarities in them.

11.7.2 Nordic Peripheries in a Network Economy

In general, the observations of this study illustrate that economic success – in this case the internationalization of SMEs – presupposes adaptation to the prevailing circumstances. A thriving enterprise is able to circumvent the constraints set by its local or regional environment. For instance, an entrepreneur in the periphery of the Nordic countries has had to accept the fact that man-made economic resources available in these regions are not comparable with those in more central locations. His locational decision has usually been motivated by personal considerations such as housing preferences.

The interviews gave the impression that small entrepreneurs are well aware of the prerequisites set by their local environment. Their assessments concerning, for instance, the quality of business and educational services and potential improvements in them were determined according to what they regarded to be economically viable in the local context. It is also interesting to observe that interpretations of a relevant local sphere of activity differed by region and by economic actor. The functional daily region of a small entrepreneur is, especially in more peripheral regions, clearly larger than the labour market district used in the delineation of the target regions in this study. Entrepreneurs in peripheries are used to acquiring services from larger centers, although it obviously causes certain extra expenses. This emphasizes the significance of the internal resources and, especially, the contact resources of a firm. In addition, it has implications for development policies: a high-level service infrastructure is obviously not possible everywhere but it should at least be provided in provincial capitals.

Although the data set of this study describes the cross-sectional situation in the early 1990s, it also includes some signals of prospective changes. An illustrative example is the fact that some entrepreneurs in small towns and rural peripheries considered their current locations would increasingly weaken their competitiveness and constrain their internationalization. Their strategic aim was to link up with the relevant resources of the industrial field by locating their activities in larger centers. This also illustrates the fact that the role of a firm in an industrial complex or cluster has an effect on its strategy (cf. Enright 1994). Another feature relevant to future prospects concerned the location of cooperative partners. The partners of the most internationalized firms were farther away, in many cases abroad. This emphasizes the need to open channels to small entrepreneurs for national and international resources. In practice, this can be done in several ways, for instance, by supporting networks which do not remain only local.

As previously noted, the industrial clusters of the Nordic economies have evolved in the context of individual countries. Their embeddedness into local and regional environments can be assumed to change under new competitive conditions. This raises the problem of the extent to which the basis of a modern industrial cluster can be geographically scattered. If specialized local enterprise networks are only possible to a limited degree in the small settlements of the sparsely populated north, the economies of these areas should be based on a network which consists of far-away settlements and transport and communications links joining them. These settlements should not necessarily be at the same level of an urban hierarchy, but contacts between economic actors operating in them can be horizontal. It is clear that the formation of a functional region of this kind presupposes good accessibility. It also implies the internationalization of the settlement system, which can be supported by the successful internationalization of enterprises.

Acknowledgements

The paper is an outcome of a comparative Nordic study on the internationalization of SMEs. In addition to the present authors, the research group consists of Poul Rind Christensen (Southern Denmark Business School, Kolding, Denmark), Bo Forsström (Institute for European Studies, Turku, Finland), Olav Jull Sørensen (Aalborg University, Aalborg, Denmark), and Eirik Vatne (SNF, Bergen, Norway). The joint data base was compiled by Timo Lautanen (Karelian Institute, University of Joensuu, Finland). The writing of this paper has greatly benefited from ideas suggested by the above mentioned colleagues.

References

Brusco, S. (1992), 'Small Firms and the Provision of Real Services', in Pyke, F. & Sengenberger, W. (eds.), *Industrial Districts and Local Economic Regeneration*, ILO, Geneva, 177-196.

Camagni, R. (1991), 'Local 'milieu', uncertainty and innovation networks. Towards a dynamic theory of economic space', in Camagni, R. (eds.), *Innovation Networks. Spatial Parspectives*, London, 121-144.

Cappellin, R. (1990), 'The European Internal Market and the Internationalization of Small and Medium-Size Enterprises', *Built Environment* 16(1), 69-84.

Christensen, P. (1991), 'The Small and Medium-Sized Exporter's Squeeze: Empirical Evidence and Model Reflections', *Entrepreneurship and Regional Development* 3, 49-65.

Christensen, P.R., Eskelinen, H., Forsström, B., Fredriksen, T. & Lindmark, L. (1991), 'Lokal resursmobilisering för internationell konkurrenskraft – några reflexioner kring småföretagsutveckling i ett EG-perspektiv', *NordREFO* 1991:1, Copenhagen, 9-28.

Christensen, P.R. and Lindmark, L. (1993) 'Location and Internationalization of Small Firms', in Lundqvist L. and Persson L.O. (eds.), *Visions and Strategies in European Integration – A North-European Perspective*, Springer-Verlag, Heidelberg, 131-151.

Commission of the European Communities. Directorate-General for Regional Policies (1993), *Impact of the Development of the Nordic Countries on Regional Development and Spatial Organisation in the Community. Final Report*, NordREFO, Copenhagen.

Davidsson, P., Lindmark, L., Olofsson, C. (1993), 'Business dynamics and differential development of economic well-being', Paper presented at Rent VII Conference, November 25-26, 1993, Budapest, Hungary.

Eskelinen, H. (1993), 'Rural Areas in the High-mobility Communications Society', in Giannopoulos G. and Gillespie A. (eds.), *Transport and Communications Innovation in Europe*, Belhaven Press, London and New York, 259-283.

Eskelinen, H., Lautanen, T. and Forsström, B. (1994), 'The Internationalization of SMEs in four Finnish regions: some tentative observations', in Lundqvist L. and Persson L.O. (eds.), *Northern Perspectives on European Integration*, NordFEFO 1994:1, 115-133.

Eskelinen, H. & Vatne, E. (1995), 'Resources or Barriers for Exporting? The significance of local networks for Nordic small and medium-sized enterprises', *Revue Internationale PME*, (forthcoming).
Enright, M.J. (1994), 'Regional Clusters and Firm Strategy', Paper presented for the Prince Bertil Symposium: The Dynamic Firm: The Role of Regions, Technology, Strategy and Organization, Stockholm, June 12–15.
Hammervoll, T. & Lensberg, T. (1992), *Et konkurransedyktig Norge. Närings- og sektoranalyse*, SNF WP 88/1992, Bergen.
Hanberger, A. (1993), *Lokalt samarbete och global integration*, ERU-rapport 77, Östersund.
Hernesniemi, H., Lammi, M. & Ylä-Anttila, P. (1995), *The Competitive Advantage of Finland*, ETLA, Helsinki, (forthcoming).
Johansson, B. (1993), *Ekonomisk dynamik i Europa. Nätverk för handel, kunskapsimport och innovationer*, Liber-Hermods, Malmö.
Kajaste, I. (1992), 'The impact of "1992" on the Finnish manufacturing industries', in *European Economic Integration; Effects of "1992" on the Manufacturing Industries of the EFTA Countries*, European Free Trade Association, Economic Affairs Department. Occasional Paper No 38. Geneva, 115–145.
Lindbeck, A. et al. (1993), 'Options for Economic and Political Reform in Sweden', *Economic Policy* 17, 220–263.
Lindmark, L. (1993), 'Specialisering och samverkan i Skellefteå', in Isaksen, A. (red.), *Spesialiserte produksjonsområder i Norden*, Nordisk Samhällsgeografisk Tidskrift, Uppsala, 89–108.
Lindmark, L., Christensen, P.R., Eskelinen, H., Forsström, B., Sørensen, O.J. & Vatne, E. (1994), *Småföretagens internationalisering – en nordisk jämförande studie*, NordREFO 1994:7, Stockholm.
Lindmark, L., Lindbergh, L. and Palm, P. (1995), *Globalization of Economic Activities and SME Development - Sweden*, OECD, Paris (forthcoming).
Lindqvist, M. (1991), *Infant Multinationals. The Internationalisation of Young Technology-Based Swedish Firms*, Stockholm School of Economics.
Lundberg, L. (1992), 'European Economic Integration and the Nordic Countries' Trade', *Journal of Common Market Studies* XXX(2), 157–173.
Miesenbock, K.J. (1988), 'Small Business and Exporting: A Literature Review', *International Small Business Journal*, 6(2), 42–61.
Porter, M. (1990), *The Competitive Advantage of Nations*, The Free Press, New York.
Räsänen, K. & Whipp, R. (1992), *Challenges of Comparative Research on National Business Recipes*, European Institute for Advances Studies in Management WP2/1992, Brussels.
Storper, M. (1995), 'The resurgence of regional economies: the region as a nexus of untraded interdependencies', *European Urban and Regional Studies*, 2(3), (forthcoming).
Sölvell, Ö., Zander, I. & Porter, M. (1991), *Advantage Sweden*, Norstedts, Stockholm.
Vatne, E. (1995), 'Local Resource Mobilization and Internationalization Strategies in Small and Medium-Sized Enterprises', *Environment & Planning A*, London, 1, 63–80.

CHAPTER 12

The Social Construction of Peripherality: the Case of Finland and the Finnish-Russian Border Area

Anssi Paasi
University of Oulu

12.1 Introduction

Once introduced and defined, the categories used in scientific discourse often become a 'taken for granted' part of scientific practice and rhetoric. In recent discussions regarding the nature of 'space' – either with material spaces, metaphoric spaces or something between – this vagueness has been a common problem. This probably results from the different concepts of space that various academic disciplines set forth now that interest in spatiality and space has increased within various disciplines (Massey 1993a). The changing conditions of movement and communication – time-space compression (Harvey 1989) – are also affecting the concepts of space and time and call for an analysis of the contents of these categories. Massey (1993b) writes about the power-geometry of time-space which suggests that different social groups and different individuals are placed in distinct relationship to these flows and interconnections. This thesis also gives rise to different representations of space.

The concept of periphery is another example of the problem of "taken for granted" spatial categories, although it is more dynamic and politically laden than many other ideas connected with space (Massey 1993b). This concept is typically employed – together with its conceptual counterpart 'centre' – in the metaphoric classification of social space according to some preselected criteria. The same term may then be used in a very different context, for example in the classification of concrete areas of earth surface. The results of these classifications may then be evaluated in terms of the politics of location, i.e. spatiality may be evaluated in political terms both in periphery and in centre.

The concept of periphery is relatively new in social scientific discourse, more recent than the idea of centre, which is more deeply rooted in the human mind and practice (Strassoldo 1980). In the rhetoric regarding the relationship between centre and periphery, the former by definition has authority power over the latter; centre is typically the place where the seat of authority is located. Space is therefore understood to constitute a continuum which is then comprehended to be divided according to certain criteria into subareas: centres and peripheries, developed and underdeveloped areas, cores and margins and so forth. Thus the concepts of centre and periphery imply both a specific relation of complementarity and a possibility for opposition and confrontation (cf. Gottmann 1980). This dialectical relationship appears upon different spatial scales.

The construction of representations of peripheries is in fact a typical illustration of what Shields (1991) labels as 'social spatialisation'. During this process, the visions of margins and cores, centres and peripheries are created on different grounds. Social spatialisation is a result of both discursive and non-discursive elements, practices and processes. It is always a blend of scientific analysis, local and non-local spatial experience, operations of media, political struggle and ideologies. These are manifested differently (on different spatial scales).

One of the most powerful and categorical classifications of horizontal spaces according to their peripheriness is provided by the World System Analysis (WSA) (Taylor 1993, cf. Strassoldo 1980). It is powerful because it has become extremely popular within many disciplines. It is categorical, because it provides a straightforward, heuristic framework for theoretical discussions and empirical research. In the terminology of WSA, 'periphery' refers to one of three major zones in the global world-economy. Whereas the 'core' areas of this system are characterised by 'core processes' involving, for example, relatively high wages and high tech production, 'periphery' is characterised by 'peripheral processes' of relatively low wage and low tech production. Between them is 'semi-periphery' which is characterised by a mix of peripheral and core production processes. The perspective of WSA may be useful in tracing main trends and the historical development of territorial economic systems at the global level, however, this framework is a relatively abstract and economistic one, and less useful with smaller spatial scales. Something else is required for the scale of state and sub-state, and even smaller socio-spatial scales, such as the internal socio-spatial differentiation of localities as well as the construction, reproduction and control of spatial and social relations between individual actors. However, because of the popularity of the WSA framework, it is common to see these categories misused in the analysis of the economic developments in smaller scales.

Centre-periphery approaches have frequently been employed in the case of economic relations. Although economic development is core to the idea of peripherality, 'periphery' is understood here more broadly to include the political, ideological and cultural aspects of the production of space, as well. As Snickars (1989) points out, both centre and periphery are complicated expressions and it is no longer the case that a geographical periphery must necessarily be an economically weak region.

In this article the concept of peripherality will be understood in a broad way in order to scrutinise the dimensions and uses of this category in various spatial scales, i.e. local, regional, national and global. Periphery may be characterised by such dimensions as distance, difference and dependence; the last of these categories typically manifests itself in political decision-making, cultural standardisation and economic life (Rokkan & Urwin 1983, 3-4). These dimensions, therefore, create different boundaries between the centre and periphery at different scales. During recent decades the world has moved from isolationism toward integration. Borderlands have become important for nation-states with significant interlinks (Martinez 1994). This is particularly relevant with regard to Europe where the collapse of the Eastern European communist regimes has changed dramatically relations among cultural, political and economic centres and peripheries.

The aim of this article is to examine specifically the roles of two spatial units and scales, the Finnish state and the Finnish-Russian border area (Karelia) as examples of the social construction of peripherality. The concept of border areas is closely connected to the idea of peripherality, since typically the closure of boundaries discourages investments, causes higher costs and has a depressing effect in general on economic investments. Open boundaries and energetic interaction in border areas attract people and capital, according to Strassoldo (1980). I will discuss the changing roles of the economic, political and cultural processes in the peripheralization of Finland and the border area, and then evaluate these processes from the viewpoint of different spatial scales. The aim is to 'map' the representations given to Finland's peripheral location at varying scales, i.e. local, national and global. My purpose is to show that in the case of Finland the idea of peripherality has had different, historically developed forms and associated representations which have varied at different spatial scales. Based on the analysis I will try to draw some conclusions that sharpen the vague idea of peripheriness.

12.2 Constructing Peripheries: Finland and the Finnish-Russian Border Area

12.2.1 Finland Between East and West

Finland is a sparsely populated nation-state in Northern Europe which is, according to several commentators, located in the 'European periphery' adjacent to the former "iron curtain" (Rokkan & Urwin 1983, Mead 1991). It may be argued that Finland is a *relative* periphery on several spatial scales: among Nordic countries, in Europe and in the global system of states. It is a cultural periphery from the viewpoint of western culture – whatever this generalisation exactly means. It has been a political periphery in the western social consciousness and geopolitical reasoning since World War II. It has also been an economic (semi-)periphery in the global system of states.

Finland has hence been a socio-spatial entity which has been located culturally, politico-ideologically and economically somewhere 'between' other entities. At first it was an emerging 'external periphery' (Rokkan & Urwin 1983, 30): an eastern part of the Swedish state, geographically remote, at the edge of western Europe and exposed to the influence of one centre. For hundreds of years Finland was a peripheral part of Sweden. In 1809 it became an autonomous state of Russia. Finally, after 1917, it became an independent part of the northern periphery of Europe (see Figure 12.1). In maps representing European core areas, Finland is typically without representation except for a small developed coastal belt (national core). Like Sweden and Norway, Finland has been characterised by a North-South economic gradient, which has at times manifested itself in regionalistic ideologies and rhetoric and, as such, has been an expression of peripherality (ibid. 42–43).

The recent efforts of the leading politicians of Finland and other Nordic states to become members of EU are an illustration of systematic aims to change the peripheral position. Economic concerns predominate, but military security is also an important issue. There is also a desire to dispel the view that Finland is a cultural vacuum. Located between the East and the West, Finland has been exploited by cultural arguments for political reasons. In present day Finland, there is an emerging emotional, somewhat nationalistic debate about possible effects of EU membership on national sovereignty. Similar national arguments are part of the debate on the possible social, political and military effects of the ongoing opening of the Finnish-Russian boundary.

It has been argued in the most extreme opinions that a decision to enter into the EU would invalidate the whole foundation of Finland's independence, whatever this literally means in the (post)modern world characterised by the compressing space of information and capital flows. Ironically, discourses about nation-building appear to be empty words in present day Finland. Economic recession and unemployment coupled with changing international economic relations are forces dividing the nation along with the apparent, gradual demolition of the welfare state. All these trends impact upon the 'civil religion', legitimation and moral landscape in social integration, one basic key-function of the state (Johnston 1989), probably for the first time since the Winter War and particularly since the Civil War in 1918.

12.2.2 Karelia as a Border Area

Boundaries may be simultaneously historical, natural, cultural, political and symbolic phenomena and each of these dimensions may be exploited in diverging ways in the construction of territoriality and peripherality (cf. Paasi 1995a). These apply in the case of the Finnish-Russian boundary and border area: located on the Karelian area where western and eastern cultures meet, its location has changed many times as a consequence of territorial disputes. Figure 12.1 shows the boundary changes since the 14th century, but in practice it is reasonable to discuss the meanings of the boundary merely beginning from the 19th century when Finnish nationalism began to emerge and especially after the year 1917 when Finland gained its independence and an explicit boundary between Finland and Soviet-Russia was established.

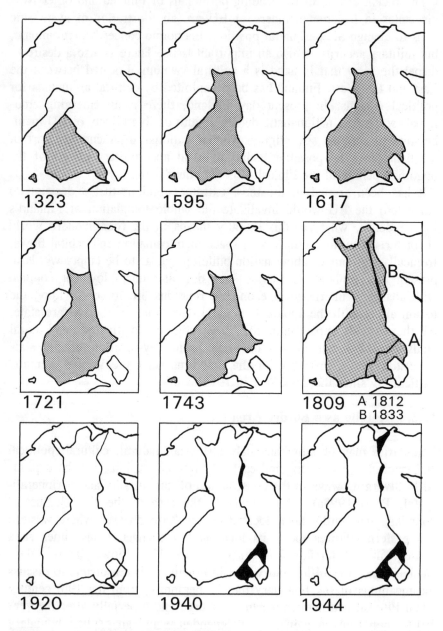

Figure 12.1. The development of the boundaries of Finland since the 14th century

Since World War II the border represented the longest ideological boundary between a western capitalist state and the Soviet Union. Finnish-Russian border areas are typically characterised by signs of national peripheriness: dependence on resource based industries, the significance of the public sector in employment, low population density, and so forth. The Finnish-Russian border area also presents an illustration of the transformation of borderlands that is taking place in the present-day Europe of territories. The Finnish-Russian boundary, and border area in general, is a particularly interesting example of the social construction of boundaries and identities and of the varying uses of the concept of peripherality in different scales.

The existence of the Finnish-Russian boundary may be understood through an historical framework which emphasises different aspects in the construction of peripherality. Four historical stages may be distinguished in this frame: 1) the period of the autonomous Grand Duchy of Russia (1809–1917), 2) the gaining of independence in 1917 and the period between the two World Wars, 3) the post-World War period extending up to the dispersion of the Soviet Union and, finally, 4) the post-Soviet period. During these periods several representations of the border, based on natural, cultural or political features, have appeared simultaneously although certain features have dominated (cf. Paasi 1995a, 1995b). In different stages varying aspects have been emphasised in the construction of the idea of peripherality both in Finland and in a broader context. The border area has been located in a national periphery (periphery understood as mainly an economic and political category), at various scales, and has been an important factor in the constitution of local peripheries both on Finnish and Russian sides of the border. One key factor behind this 'peripheralization' has been the Cold War which, after World War II, created a strict division between east and west. Now the situation is changing dramatically as a consequence of the political and economic changes that have taken place in the former eastern Europe. This process is changing once again the understanding of the peripherality of these areas.

12.2.3 Nationalising the Finnish Peripheries

The production of national identities and boundaries may be regarded as part of a project in which physical and symbolic construction and reproduction of the territories and their internal social integration takes place

(Paasi 1995a, b). The ideological foundation for this process is most typically nationalism, through which the physical and political territory of a nation is transformed and expressed symbolically as a cultural entity. It has been argued that nationalism is two-faced with respect to space (Anderson 1986). It looks inwards in order to unify the nation and its constituent territory and outwards to divide one nation and territory from another. Williams and Smith (1983) propose that nationalism is always concerned with struggle over the control of land. Nationalism both constructs and interprets social space. The process of 'nation-building' aims toward incorporating the state and its inhabitants (Deutsch 1963; MacLaughlin 1986). Key media in nation-building are the socialisation mechanisms controlled by the state. Particularly significant is the education system, through which new generations can be incorporated as members of a national community (Paasi 1991, 1995a).

Finland's achievement of independence in 1917 caused a change in its territorial strategy: it had to secure its boundaries and thus signify the territoriality of the state. The boundary established in the Peace of Tartu (1920) was located in the same place as the boundary of Grand Duchy of Finland and Russia but in practice, the meaning of the new boundary was completely different from the old one. State power was established in border areas in the form of a border guarding system. Whereas it was possible prior to 1917 to interact and communicate over the boundary, after the establishment of independence this was no longer possible; the new boundary was in principle closed. In reality, illegal interaction over the boundary took place which also caused diplomatic conflicts (Paasi 1995a, 1995b).

As a whole the closure of the boundary was a radical change for the less developed areas in eastern and east-western Finland. It promoted their economic peripheralization and changed their symbolic and cultural roles as the outpost of the western cultural realm. Commercial connections to the east were broken. Particularly connections with the significant economic sphere of influence and market area around St. Petersburg were severed from the Finnish side. The border was nearly closed and a minimal amount of interaction over the boundary took place. Control of the border area was maintained through effective control mechanisms both at the scale of the Finnish state, peripheral border regions and at the local scale and daily life of the individuals and groups living in the border areas.

After the Finnish state gained its independence, one of the most important national concerns become securing the independence. Activities took place particularly to develop the peripheral areas located near the Soviet, but also in the Norwegian and Swedish borderlands. Agricultural activities in these areas were also supported by the State beginning in 1923 (Paasi 1995b). These aims where obvious expressions of efforts towards 'nationalising' the peripheries, i.e. the government's aims to replace vague frontiers by sharp boundary lines and to integrate the peripheral borderlands with the rest of the core area of the state (cf. Augelli 1980; Kliot and Watermann 1983). In principle this can occur in three ways: in efforts to strengthen internal integration and identity, to lessen external influences or by exploiting both of these strategies.

In Finland the territorial motives to develop the economic living conditions in border areas were originally based upon the idea of promoting national level social integration. This involved ideologically-laden efforts to eliminate the distinctions between social classes: rich and poor, owners and workers. At the same time, however, a clear external demarcation was made at the Soviet Union. This contributed to the creation of an eastern border area periphery inside the state (cf. Rokkan & Urwin 1983). The territorial aim of the boundary reinforcement in the 1920s and 1930s was the national integration within Finland. Great efforts were made toward developing the peripheral border areas and improving the living conditions of their inhabitants. These were intended to foster a spirit of nationalism and to increase the political reliability of the population. While these activities present a good expression of the nationalisation of peripheries, at the same time it was difficult for the local people to comprehend why they were 'relocated' into a periphery and why their rights which once extended fairly freely across the border now had to be curtailed (Paasi 1995b).

In addition, during the 1920s, the specific role of the eastern boundary itself and the areas locating near the boundary become important in civil activities. Immonen (1987, 317) concludes that the boundary became the symbol through which both the contradistinction between Finland and Russia/Soviet Union as well as their dependence upon each other was expressed. The meanings of the boundary were hence effectively politicised. In this sense, thinking on both sides of the border was similar. However, the Soviets did not view the western boundary of their country as a "natural one": rather, every neighbour beyond the border was a potential enemy, since the governments of neighbour states were clearly anti-Moscow. For the Soviet authorities, Finland was the Northernmost, geopolitically important link in this chain of enemies. It was regarded as aggressive and anti-Soviet (Korhonen 1966, 30–31).

Therefore, it can be argued that, with the rise of the Socialist Soviet Russia and Finnish independence, the meaning of the border between Finland and Russia was changed from a strictly legal one to a clearly symbolic and closed political boundary (Hämynen 1993). For the new independent state of Finland this demarcation was an essential part of the 'nation-building' process of the new state (Paasi 1995b).

12.2.4 The Cultural Periphery of the West?

One feature that emerged as a result of independence was the shaping of a *cultural* boundary between Finland and Soviet Union. Although it had been common to emphasise the eastern historical roots of the Finnish people, particularly linguistically, there now emerged an acute need to construct an image of Finland as a western country. Thus Finland was often represented as the last peripheral outpost of western Europe. The argument given for this representation was usually Finland's long connections with Sweden and Central Europe. The purpose was to present the Finnish area as clearly as possible as a part of western civilisation instead of as a mere cultural vacuum adjacent to the former mother state (Soviet-)Russia (Paasi 1995a, 1995b).

Language is one of the key constituents of a cultural identity. It has always been one of the central dimensions that has made Finland a periphery on the European and even Nordic scale (see Väyrynen 1985, 53). The Finnish language has its origins in the Ural area and this inevitably results in an eastern flavour to Finnish culture and "Finnishness". Language has also been a key factor in the ethnic identity of the Finns. The Finnish language has dominated while the Swedish speaking minority has been located almost totally in the western coastal areas of the country. This, in a way, forms a transformation area in the larger cultural space of the Nordic countries. Even if the distance between the peripheral (Finnish) and the former colonial elite language (Swedish) has always been much greater in Finland than in many other countries (Rokkan & Urwin 1983), it can be argued that the Swedish-speaking area has been an important sphere of mediation in the construction of the Finnish national identity, since it is partly based on separation from the eastern cultural space. The Swedish language is also key in the integration of Finland into the cultural and political co-operation of Nordic countries. With these features in mind, Finland can certainly be viewed as part of the western world. This statement has also been used historically in defence of Finland's location on the global geopolitical scene.

During the 1920s and 1930s the intensifying physical and psychological separation from the new Soviet Union was one part of a general international geopolitical trend. The eastern border gradually became a mythical and symbolic expression of a historical and evidently eternal opposition between two states: Finland was understood as the last bastion of Christianity and western ideals (Paasi 1990). This position was actively inculcated into the collective memory of succeeding generations, mainly through education in the schools.

Because of its location between the Eastern and Western blocks and the cultural influences that have come during hundreds of years from both directions, the border area has been, from the beginning of independence up to recent years, laden with strong ideological connotations and it has been one essential constituent in the iconography of the cultural and political 'Finnishness' and in the creation of the internal and external representations of Finland's place in the world. Particularly in the time just before World War II and during the war years, some Finnish authors tried to classify Finland strictly and exclusively in the realm of Western Europe and Western cultural heritage – even if the eastern connections have at times been acknowledged as a background for these developments (Paasivirta 1992, 70, 206–207. This ideological stance also manifested itself in writings of several Finnish scientists in the 1920s and 1930s as well as in school geography textbooks (Paasi 1995b).

12.2.5 Locating on the Global Geopolitical Periphery

The new global geopolitical order that developed after World War II meant remarkable change with regard to the territorial structures and the centre-periphery relations of Europe. A new world-wide spatial scale could be distinguished in the context of international conflicts as a result of the east-west dichotomy. The world separated out into three spheres: 'ours', 'theirs', and a set of disputed areas which had no obvious 'owner' or reference group (cf. Gottmann 1973; Taylor 1988). Accordingly, this also created new cores and peripheries in the global geopolitical system.

From the point of view of world geopolitics, Finland became situated in the disputed, indeterminately neutral camp, located somewhere between East and West. After World War II Finland's eastern boundary became the longest one between the leading socialist state and a Western capitalist state, over 1,200 km long. It also provided a fitting illustration of an *ideological boundary,* at times quoted in political geography textbooks (see Pounds 1972, 99; Vuoristo 1979, 55–56).

The effects of the new boundary differed from the previous one. Along with the changing position of the boundary, Finland lost certain essential territorial, cultural and historical elements of her eastern character and of the eastern legacy that helped create her identity. The loss of the ceded areas also created a severe economic problem for an agrarian slowly industrialising country, with 432 factories, over 25,000 industrial jobs and 25 per cent of the country's hydro-electric capacity having been left in the ceded areas, together with prominent towns such as Vyborg and Sortavala. Furthermore, Finland had to produce as war indemnities various goods, especially metal and timber products, for the Soviet Union to a final value totalling about 450–500 million dollars, amounting to between 10 and 20 per cent of national income. At the local scale the new boundary once again meant the economic peripheralization of the areas in eastern Finland (Paasi 1995b).

The war indemnities and the return of all evacuated machines, locomotives, motor vehicles etc. also marked the beginning of a new economic interaction between the two states after the war. The trade agreement for the period 1951–1955 became a basis for continuing bilateral trade. With regard to timber, pulp and paper exports, Finland was linked to the western European economy. Soon after the war the Soviet Union and Britain became Finland's most important partners in trade, while until 1955 Britain was the most important partner for Soviet trade and Finland was the second (Heikkilä 1986; Paasi 1995a).

On the national scale, the eastern political presence in Finnish society was strengthened by the official agreements made with the Soviet Union. In Finland, the boundary became a symbolic boundary between east and west, between rich and poor, between democracy and totalitarianism (Paasi 1995a). Finland's new geopolitical situation also directed activities in foreign policy. It became common to see the Finnish case as unique, particularly when compared with other countries in the

"Soviet orbit" (cf. Vloyantes 1975; Allison 1985). This position effected the representations of this position, which can be seen in political, geographical depictions of Finland's position on the world map.

The dichotomy between East and West that was established as a consequence of the geopolitical division after World War II, was based on political and economic arguments. Its basic idea was the division between capitalist and socialist states. As far as the foreign representations of Finland's place in the world are regarded, it is interesting to note that in some Anglo-Saxon textbooks on political geography, Finland after the war was located to the eastern realm by some unstated arguments, obviously partly due to its physical location and partly on the grounds of its cultural or political connections. Before the war it was usually located in Western Europe. In Alexander's (1957) "World Political Patterns", for instance, Finland has been located to eastern Europe. According to the author it is "something of an anomaly among eastern European countries, since it has not been occupied and since World War II it had maintained as neutral a position as possible with respect to the Cold War".

In Cohen's (1964) well-known frame of geo-strategic regions, Finland's territory is mostly located in Eastern Europe. Further, Cohen labelled Finland, together with Austria and Yugoslavia as 'buffer' states. Cohen's schema was very much an American one (Taylor 1988). It was based on Finland's obvious physical location but perhaps also on his interpretation of Finland's economic and political connections with the Soviet Union. Particularly the Treaty of Friendship, Co-operation and Mutual Assistance served to locate Finland in the Soviet sphere of interest.

Interestingly enough, Cohen (1991) has revised his cartographical scheme of the geo-strategic regions of the world and Finland, together with the former eastern European states (e.g., the Baltic states), is now located in an intermediary area between west and east. Finland is still considered an 'anomaly'. It is no more a part of the east as in Cohen's (1964) former study but neither is it part of the west, as leading politicians of Finland claimed in their efforts to enter the EU. This is an illustration of a typical difficulty in changing relatively fixed spatial representations in different spatial scales and also an example of the symbolic construction of space and peripheries.

Before the radical changes in eastern Europe, Väyrynen (1985) wrote that Finland has been both geopolitically and economically between East and West. Geopolitically it was placed between two military blocks, Nato and the Warsaw Pact; economically Finland also traded extensively with former socialist countries. Väyrynen characterised Finland as having semi-peripheral features in its international position. Characteristic of this situation was the heavy state control of industrial production. Whereas the border areas on both sides of the boundary were located in an economic periphery, at the national scale Finnish-Soviet economic relations developed rapidly after World War II. As stated earlier, the Soviet Union and Britain became Finland's most important partners in trade after the war (Heikkilä 1986).

12.2.6 From Peripheral Areas to Fields of Interaction

The new boundary with Finland and the territories that Finland had to cede after World War II were also objects of redefinition in the Soviet Union (Paasi 1995a, 1995b). Whereas in Finland the basic idea was to nationalise the peripheral areas of the territory, one could argue that in Soviet Karelia the idea was to peripheralize and non-nationalise the space after World War II. This redefinition took place ideologically but also in the form of concrete activities, such as a settlement policy which sought to establish the dominating population of Russian Karelia as non-Karelian people. In the most extreme circumstances of the war years, commanders of the Soviet Army suggested that the area should be emptied of all its inhabitants. However Stalin had other plans (Paasi 1995a). Accordingly, people from Russia, the Ukraine and White Russia began to settle the area (Laine 1994). Since the resettled people had entered from different parts of Soviet Union, they surely had no common historical feeling of belonging to these areas.

The end of the Cold War has given rise to a new economic growth and restructuring for many border areas located between eastern and western Europe, and the formerly closed borders have rapidly developed into *spheres of contact* between countries (Paasi 1995a, 1995b). In many areas of Europe co-operation over boundaries has become a common feature and in some places (e.g., Basel region) we may even talk about 'integrated borderlands' (cf. Martinez 1994). The key actors in these areas perceive the areal entity as one social system, economies of different countries may be functionally merged and goods and people may move across the boundary almost without restrictions.

Although there has been extensive trade between Finland and Soviet Union until recent years, seventy years of almost no transborder activity has made the peripheral areas on both sides dependent upon their own national political and economic centres (Paasi 1995a). The areas involved could be labelled 'alienated borderlands': the border was functionally closed and cross-border interaction was nearly absent with the exception of some strictly controlled points of contact (cf. Martinez 1994). In Finland the aim of regional policy has been to keep the peripheral rural areas as well as border areas inhabited by providing economic support for them. In spite of this fact, the border areas in eastern Finland have lost much of their population through migration during the last decades, mainly due to the diminished economic possibilities and increased unemployment. On the other side of the border, as a consequence of conscious investment and industrial policy, the Russian Karelian area urbanised rapidly after the war and accordingly only a minority of the population lives in proper rural areas (Nevalainen 1993; Varis 1993).

The collapse of the communist regime has changed the economic, political and military landscapes of Europe during the last few years. Changes are also occurring in the peripheral areas near the Finnish-Russian boundary. The political and economic changes in eastern Europe have been crucial as far as the most recent idea of the Finnish-Russian border is concerned. The interpretation of the roles of the boundary and emerging economic practices are in a new stage and it can be argued that now the idea of the border is developing into an interface, a contact surface and perhaps even to a frontier with a new social and economic significance based upon personal interaction between traders. Its economic implications are now being increasingly realised on both sides (Joenniemi 1994). The border area is in fact expanding and the forms of interaction are expanding. These areas are becoming 'interdependent borderlands', to employ the terms of Martinez (1994).

Many of the local authorities on the Finnish and Russian sides have been ready to play an active role, hoping to open up routes and connections in the future and thereby develop the economy of both areas. Although the interaction across the boundary has increased markedly, it remains a fact that the border represents a line between two completely different societies and the gap between the standards of living prevailing on each side is among the largest in the world. It is therefore likely that the Finnish-Russian boundary will be subject to increasingly strict controls, on both sides, in the near future (Paasi 1995a, 1995b). It is improbable that these areas will become 'integrated borderlands' in the near future, where the unrestricted movement of people, goods and ideas will prevail.

Nevertheless administrators and entrepreneurs in many communes located along the eastern border are optimistic that they can change their peripheral locations and open communications with Russian areas. There exist several new border stations and passport checkpoints in the border area. Five of these are official stations where it is possible to cross the border for various purposes, to promote cultural exchange, group tourism, economic activity and friendship. Two temporary checkpoints allow border crossings in order to diversify economic activities in adjacent areas (Ervasti 1994). The communes of the border area are looking forward to a future in which this will become a vital international boundary in Europe.

All Finnish circles are not so delighted with recent developments, however. The Finnish military leadership, for instance, has worried about the strategic changes in military geography that will probably take place after new road connections have been built over the Finnish-Russian border, particularly in Northern Finland. It seems that no explicit threat from Russia is experienced in military circles, but old images of "the enemy" still exist, although not explicitly directed toward Russia (cf. Joenniemi 1993). One future theme in this search for a new regional balance will certainly be the changing attitudes of the Finns towards integration into Western Europe and the possible forms of West European co-operation in the fields of economics, politics and military activities.

12.3 Living in the Forest Periphery: the Perspective of Local Experience

The building of national identities and 'imagined communities' (Anderson 1991) are always expressions of a general socialisation process that is taking place in a society. On the local scale and in daily life, the creation of identities, historical meanings, culture and heritage become much more complicated, fragmented and diversified (Paasi 1995a). It may be argued that local identities form a continuum, since memory is furnished not only from the recollections of events which an individual had experienced but also from memories of older generations. Shils (1981) argues that individual histories always include elements of the history of a 'larger self': family, neighbourhood, locality and nationality, for instance. This collective history unites the individual with the histories of these broader entities. Similarly the question of peripherality becomes very concrete at local scale: distances, forms of well-being, cultural, economic and administrative practices are lived and experienced, they are no longer forms of formal, external classification. For instance, in the border areas the social control of people (at the scale of 'body') in relation to the frontier zone is typically very strict. This has also been the situation on the Finnish-Russian border areas, both on the Finnish and Russian side.

As a consequence of World War II Finland had to resettle more than 420,000 inhabitants from the areas that were ceded to the Soviet Union. With the exception of just a few places, it was impossible even to visit the ceded territories of Karelia. The loss of the areas was therefore a serious and traumatic problem for the socio-spatial identity of the resettled people and the formation of new ties and cultural adaptation to new socio-spatial frames. The Karelians preserved their lost landscapes and homes in their collective memory through literature, collective action, myths, etc. The return to home and the golden past became a major goal for most of them. The ceded Karelian areas became a land of dreams (Paasi 1995a).

The attitudes towards the Finnish-Russian boundary seem to vary widely among different generations living in Finland. Those generations who have experienced their old communities and the creation of the new border and the loss of their home locality seem to live in a 'utopian world' where the memories of the old, lost community still form an essential part of their regional identity. However, the generations who have not experienced the war, live their daily life in a context which has always been limited by certain geopolitical facts. For them the boundary has always been where it is now. Similarly, the rules and practices of this area have been part of daily routines. They have no experience of different situations and former territorial disputes. There also exists another important difference between the generations. Whereas older people feared the new border after the war, younger generations have had a more neutral relationship, as it has been a natural part of their everyday life (Paasi 1995a, 1995b).

After World War II the significance of the eastern roots of the Finnish culture gradually began to come to light once more, to a great extent as a consequence of the redefinition of cultural and economic peripherality. From the 1970's onwards, the active role of the Karelian societies, the efforts of a few local authorities in the communes of Eastern Finland and the general expansion of the tourist industry have served to create a 'new Karelia'. The eastern flavour involves a set of images arising from Karelianism and Orthodox religion just as much as from its high-class hotels and restaurants, summer festivals and monuments from the last war. The 'new Karelia' has succeeded in providing resettled Karelian refugees with mental images which can substitute for their lost territory, just as it has stimulated business in the tourist sector in the peripheral areas of eastern Finland. Schemes for building houses in the Karelian style reinforce their commercial interests through a nostalgic desire to recreate on the Finnish side of the border something of the lost territory (Paasi 1995a, 1995b).

As suggested above, the Finnish-Soviet boundary was often represented during the existence of the Soviet Union as an ideological boundary between west and east, the capitalist and socialist worlds. The boundary was also exploited in the field of tourism as the cultural boundary between west and east, in a way, as the extreme limit of the periphery of the western world. As regards the foreigners and people from western and southern Finland, the boundary was a mysterious place during the existence of Soviet Union. The division between east and west was and still is less visible among local people, whether they belong to older or younger generations. This is due to the fact that for them the border is an essential part of the practices of daily life and not merely a representation reflecting wider historically constructed sociospatial consciousness prevailing in Finnish society.

The dispersion of the Soviet Union finally opened Russian Karelian areas to tourism. There are still about 180,000 of the Finnish Karelian refugees alive today, and an immediate boom in nostalgic journeys to Karelia has ensued, with a total of 1.26 million crossings of the Finnish-Russian border in 1991–1992 (Lehto & Timonen 1993). Probably every Finn is now familiar with TV-programmes and newspaper photographs showing former refugees searching the fields and forests of Karelia for their lost homes and past spatial identities which were broken off as a consequence of the war. These have been visits into the past as much as journeys undertaken in the present (Paasi 1995a).

There exists still one important territorial theme in recent Finnish debates that is significant as far as the future of the peripheral areas in eastern Finland and Russian Karelia are concerned. The collapse of the Soviet Union has given rise to discussions of the 'historically justifiable' boundaries in Finland and of the possibility of areas ceded at the end of World War II being returned to Finland. On the other hand, also in Russia some expansionist conceptions of Russian territory have been put forward, especially by the extreme-nationalist Vladimir Zhirinovsky. In Finland this question has been raised most actively by representatives of the Karelian Societies and it is more a national question than a local one. According to some, Finland should get the ceded areas back but there has also been some critical discussion about what should happen to these regions in a possible Finnish connection: what would be the costs of building up a new infrastructure within these peripheral areas, what would the future be of the huge ethnic Russian minority, and so forth. No support for such conjectures, however, can be derived from official Finnish foreign policy. Official opinion in Finland coincides with that expressed by the Russian authorities. On this line the former peripheral border areas will benefit by the increasing interaction between economic actors. This will probably also change their peripheral nature in the future.

12.4 Conclusions: Towards a Contextualized Concept of Peripherality

On the basis of the previous conceptual and empirical analysis of Finnish-Russian border areas, I will now draw some conclusions in theoretical terms in order to unravel the complicated idea of peripherality. I will emphasise particular viewpoints which have special relevance to peripherality.

It is clear that peripherality is, or should be, a *contextual category* rather than a mere technical instrument to be employed in the classification of social – mainly economic – space. This contextuality can facilitate an understanding of the cultural production of space, and the production of cultural meanings and significations, which may then be exploited economically, politically or ideologically. Marginal areas, then, may not be merely geographical peripheries but also cultural, political or ideological peripheries in the cultural systems of space (cf. Strassoldo 1980, Shields 1991). On the other hand, it should be noted that geographical peripheries are not necessarily economic or cultural peripheries (cf. Snickars 1989).

Peripherality is typically connected with the horizontal division of space. What is also needed, is the identification of what may be called 'the scales of peripherialization'. In recent years several geographers have emphasised the importance of scale in the production and reproduction of space, as well as the representations of space (Taylor 1982, 1993; Smith 1993). It may be argued that whereas 'space makes a difference', so also does scale. Different scales may be characterised by different mechanisms of power and control. For instance, the power mechanisms exploited in the control of the scale body, or local, regional or national power systems may be characterised by different forms.

Centre-periphery systems and the processes of marginalization exist simultaneously in different scales of territorial structure, extending from the scale of body to the global scale. This means that peripherality should be a contextual but is also a relativistic category; there are no 'absolute' peripheries. The changing discourse on Finland's position among western/eastern realms and the representations put forward by political geographers are evidence of this relativism. They are also fitting illustrations of Foucault's (1980) argument, according to which each society has its own regime of truth, the types of discourse which it accepts and makes function as true. This accentuates the fact that texts have to be interpreted within their contexts – this applies particularly well to representations of the political order of the world, including the construction of peripheries.

The idea of scales of peripherialization is reminiscent of the idea of 'vertical peripherality', discussed by Rokkan and Urwin (1983). This concept, however, still points to the existence of two basic spatial units, the 'centre' and 'periphery', whose groups of actors may then be classified (e.g., according to their power). The scales of peripherialization point to the fact that various centres of power and margins and marginalities always exist simultaneously at various scales, beginning with local, personal human relations and extending to the major ideological geopolitical divides of space.

Peripheries are produced and reproduced in the course of complicated economic, political, cultural and social processes taking place on various spatial scales. Peripheries mean different things in different places and for different people. Hence, for instance, the abstract dichotomy between core, semi-periphery and periphery has mainly heuristic value in the depiction of centre periphery relations. In fact, we may well speak about the peripheries and cores of *knowledge* in the production of representations (Chambers 1983); the representations of the peripheries are typically constructed and defined in cores. Peripheries are particularly important as metaphors of space, but not only as examples of material space, since cultural meanings of space may be significant (e.g. in the construction of the economic meanings of peripheral areas).

Since the territorial system is always changing, the analysis of peripheries cannot ignore the historical context. Peripheralization is an ongoing process and different forms of peripheries and divergent processes of peripheralization can be understood merely historically (cf. Gottmann 1980).

We often tend to think in terms of a centre-periphery system and emphasise the relations between these two socio-spatial entities. In the present day world more emphasis should be placed upon the contact or border areas between different centre-periphery systems. This is particularly important in the case of the European system of states, in which traditional border areas are more and more often becoming areas of active interaction.

In conclusion, it may be argued that the usefulness of the peripherality concept depends upon our ability to link this abstraction to concrete social, economic, cultural and political processes that take place or manifest themselves in various spatial contexts.

References

Alexander, L.M. (1957), *World Political Patterns*, John Murray, London.
Allison, R. (1985), *Finland's Relations with the Soviet Union*, St.Antony/Macmillan, London.
Anderson, J. (1986), 'On Theories of Nationalism and the Size of States', *Antipode*, 18, 218–232.
Anderson, B. (1991), *Imagined Communities: Reflections on the Origin and Spread of Nationalism*, Verso, London.
Augelli, J. (1980), 'Nationalization of Dominican Borderlands', *Geographical Review*, 70, 19–35.
Chambers, R. (1983), *Rural Development. Putting the Last First*, Longman, London.
Cohen, S.B. (1964), *Geography and Politics in a Divided World*, Methuen, London.

Cohen, S.B. (1991), 'Global geopolitical change in the post-Cold War era', *Annals of the Association of American Geographers*, 81, 551–580.
Deutsch, K.W. (1963), 'Some problems in the Study of Nation-Building', In Deutsch, K. and W.J. Foltz (eds.), *Nation-Building*, Atherton Press, New York.
Ervasti, O. (1994), *Kelloselkä - Kuolan portti: itärajan avautumiseen kohdistuvat odotukset ja asenteet Sallassa ja Kemijärvellä 1992*, Unpubl. M.Sc-thesis, University of Oulu, Department of Geography.
Foucault, M. (1980), 'Truth and Power', In Gordon, C. (ed.), *Michel Foucault, Power and Knowledge, Selected Writings 1972–1977*, Pantheon Books, New York.
Gottmann, J. (1973), *The Significance of Territory*, The University Press of Virginia, Charlottesville.
Gottmann, J. (1980), 'Confronting centre and periphery', In Gottmann, J. (ed.), *Centre and Periphery. Spatial Variation in Politics*, Sage Publications, London.
Harvey, D. (1989), *The Condition of Postmodernity*, Basil Blackwell, Oxford.
Heikkilä, H. (1986), *Neuvostoliiton ja Suomen väliset taloussuhteet 1945–55*, Historiallinen Arkisto, 88, 109–144.
Hämynen, T. (1993), *Liikkeellä leivän tähden: Raja-Karjalan väestö ja sen toimeentulo 1880–1940*, Historiallisia tutkimuksia 170, Tampere.
Immonen, K. (1987), *Ryssästä saa puhua... Neuvostoliitto suomalaisessa julkisuudessa ja kirjat julkisuuden muotona 1918–1939*, Otava, Helsinki.
Joenniemi, P. (1993), 'Euro-Suomi: rajalla, rajojen välissä vai rajaton', In Alapuro, R. and K. Pekonen (eds.) *Suomesta EuroSuomeen. Keitä me olemme ja mihin matkalla*, Rauhan- ja konfliktintutkimuslaitos, tutkimustiedote No. 53, Tampere.
Joenniemi, P. (1994), *Regionality and the modernist script; tuning into the unexpected in international politics*, Occasional Papers, No. 57, Tampere Peace Research Institute.
Johnston, R.J. (1989), 'The State, Political Geography, and Geography', In Peet, R. and N. Thrift (eds.), *New Models in Geography*, Vol. I, Unwin Hyman, London.
Kliot, N. & S. Watermann (1983), 'Introduction', In Kliot, N. and S. Watermann (eds.), *Pluralism and Political Geography: People, Territory and State*, Croom Helm, London.
Korhonen, K. (1966), *Naapurit vastoin tahtoaan. Suomi Neuvosto-diplomatiassa Tartosta talvisotaan I (1920-1932)*, Tammi, Helsinki.
Laine, A. (1994), 'Karelia between two Socio-Cultural Systems', In Eskelinen, H., J. Oksa and D. Austin (eds.), *Russian Karelia in Search of a New Role*, Karelian Institute, University of Joensuu, Joensuu.
Lehto, T. & S. Timonen (1993), 'Kertomus matkasta kotiin. Karjalaiset vieraina omilla maillaan', *Kalevalaseuran vuosikirja*, 72, 88–105.
MacLaughlin, J. (1986), 'The Political Geography of 'Nation-Building' and Nationalism in Social Sciences: structural vs. dialectical accounts', *Political Geography Quarterly*, 5, 299–329.
Martinez, O.J. (1994), 'The Dynamics of Border Interaction. New Approaches to Border Analysis', In Schofield, C. H. (ed.), *Global Boundaries, World Boundaries*, Vol. 1, Routledge, London and New York.

Massey, D. (1993a), Politics and Space/Time. In Keith, M. and S. Pile (eds.), *Place and the Politics of Identity*, Routledge, London and New York.
Massey, D. (1993b), 'Power-Geometry and a Progressive Sense of Place', In Bird, J. et al. (eds.), *Mapping Futures. Local Cultures, Global Change*, Routledge, London and New York.
Mead, W.R. (1991), 'Finland in a Changing Europe', *Geographical Journal*, 157, 307–315.
Nevalainen, P. (1993), 'Karjala 1900-luvulla', *Terra*, 105, 316–323.
Paasi, A. (1990), 'The Rise and Fall of Finnish Geopolitics', *Political Geography Quarterly*, 9, 53–65.
Paasi, A. (1991), 'Deconstructing regions: notes on the scales of spatial life', *Environment and Planning* A, 23, 239–256.
Paasi, A. (1995a), 'Constructing Territories, Boundaries and Regional Identities', In Forsberg, T. (ed.), *Contested Territory: Border Disputes at the Edge of the Former Soviet Empire*, Edward Elgar (forthcoming).
Paasi, A. (1995b), *Territories, Boundaries and Consciousness. The Changing Representations of the Finnish-Russian Border*, John Wiley (forthcoming).
Paasivirta, J. (1992), *Suomi ja Eurooppa 1939–1956*, Kirjayhtymä, Helsinki.
Pounds, N. G. J. (1972), *Political Geography*, McGraw-Hill Book Company, New York.
Rokkan, S. & D. W. Urwin (1983), *Economy, Territory, Identity. Politics of West European Peripheries*, Sage, London.
Shields, R. (1991), *Places on the Margin. Alternative Geographies of the Modernity*, Routledge, London and New York.
Shils, E. (1981), *Tradition*, The University of Chicago Press, Chicago.
Smith, N. (1993), 'Homeless/global: Scaling Places', In Bird, J. et al. (eds.), *Mapping the Futures. Local Cultures, Global Change*, Routledge, London and New York.
Snickars, F. (1989), 'On Cores and Peripheries in the Network Economy', *NordREFO* 1989:3, 23–35.
Strassoldo, R. (1980), 'Centre-periphery and system-boundary: culturological perspectives', In Gottmann, J. (ed.) *Centre and Periphery. Spatial Variation in Politics*, Sage Publications, London.
Taylor, P. J. (1982), 'A Materialist Framework for Political Geography', *Transactions of the Institute of British Geographers*, New Series 7, 15–34.
Taylor, P.J. (1988), *Geopolitics revived*, University of Newcastle upon Tyne, Department of Geography, Seminar Papers, Number 53, Newcastle.
Taylor, P. J. (1993), *Political Geography: World-Economy, Nation-State and Locality*, Longman, London.
Varis, E. (1994), 'Gridino ja Virma – kaksi Karjalan kylää', *Terra*, 105, 316–323.
Vloyantes, J. P. (1975), *Silk Glove Hegemony. Finnish-Soviet Relations 1944-74: A Case Study of the Theory of the Soft Sphere Influence*, The Kent State University Press, Ohio.
Vuoristo, K.-V. (1979), *Poliittiset ja taloudelliset alueet*, Gaudeamus, Helsinki.

Väyrynen, R. (1985), 'Small States and Technological Dependence: Austria and Finland Compared', In Alapuro, R. et al. (ed.), *Small States in Comparative Perspective*, Norwegian University Press, Norway.

Williams, C. & A. D. Smith (1983), 'The National Construction of Social Space', *Progress in Human Geography*, 7, 502–518.

Author Index

Alber, J. 159
Alexander, L.M. 247
Allison, R. 247
Amin, A. 11, 101, 105–106, 110–111
Anderson, B. 250
Anderson, J. 242
Anderson, J.J. 83
Anderson, M. 63, 76, 80
Appadurai, A. 66
Armstrong, H. 82, 84, 86, 88
Asheim, B. T. 106
Augelli, J. 243
Axelsson, S. 152
Batey, P.W.J. 43, 48, 50
Begg, I. 85, 87
Berglund, S. 157, 179
Beynon, H. 105–106
Boadway, R.W. 78
Brunet, R. 36
Brusco, S. 207
Caciagli, M. 97
Camagni, R. 105, 110, 207
Capello, R. 20
Capineri, C. 28
Cappellin, R. 7, 41, 43–44, 47–48, 50, 54, 77–78, 215
Carlsson, F. 108, 171
Casella, A. 81
Castells, M. 161
Cecchini, P. 117
Chambers, R. 255
Charpentier J. 102
Christensen, P.R. 210–211, 215, 232
Cohen, S.B. 247
Cooke, P. 103, 105
Cox, K.R. 105
Cronberg, T. 193
Davidsson, P. 205
Dellenbrant, J. Å. 110
Denters, B. 79
Deutsch, K.W. 242
Dodgshon, R. A. 184
Dréze J. 103
Edin, P.A. 167

Eklund, K. 157
Ekman, A.-K. 107
Engel, C. 102
Enright, M.J. 231
Erlandsson, U. 116
Ervasti, O. 250
Eskelinen, H. 11, 110–111, 209, 211, 232
Esping-Andersen, G. 159
Flynn, A. 202
Forsström, B. 232–233
Foss, O. 119
Foucault, M. 254
Fournier, S. F. 10
Fredriksen, T. 232
Frey, B. 81
Froebel, F. 64
Galtung, J. 184
Ginsburg, N. 153, 158
Goldsmith, M. 102, 105
Gottmann, J. 236, 245, 255
Gupta, A. 62
Haass, J.M. 125, 128
Hall, D. 19
Hammervoll, T. 208
Hanberger, A. 225
Hansen, N. 76, 91
Harvey, D. 105, 235
Hebbert, M. 102, 108
Heikkilä, H. 246, 248
Heinrichs, J. 74
Hernesniemi, H. 208
Herod, A. 61
Hitiris, T. 77, 83–84
Hobsbawm, E. 62
Holmberg, S. 158
Holmlund, B. 167
Hudson, R. 105–106
Häkkilä, M. 202
Hämynen, T. 244
Illeris, S. 10, 120, 130
Immonen, K. 243
Imset, Ø. 107–108
Isaksen, A. 105, 143

Joenniemi, P. 249–250
Johansson, B. 127, 207
Johansson, M. 112
Johnston, R.J. 239
Jonas, A. 60–61
Jussila, H. 110, 202
Kajaste, I. 209
Kamann, D.J. 20
Karlqvist, A. 3
Kearns, G. 105
Keating, M. 103
Kliot, N. 243
Knowles, R. 20
Kokkonen, M. 10, 97
Kolehmainen, E. 202
Korhonen, K. 244
Koskinen, T. 186
Kreye, O. 74
Krugman, P. 6
Kunzmann, K.R. 104
Laine, A. 248
Lammi, M. 233
Lautanen, T. 232
Leavitt, J. 62
Lehikoinen, A. 202
Lehto, T. 252
Lensberg, T. 208
Lever W.F. 103
Lindbeck, A. 209
Lindbergh, L. 212–213
Lindell, C. 130
Lindmark, L. 11, 205, 211–215, 220, 229, 232
Lindqvist, M. 215
Lowe, P. 202
Lundberg, L. 208
MacLaughlin, J. 242
Maggi, R. 39
Maillat, D. 50
Malinen, P. 192
Malmberg, A. 11, 106
Mandel, E. 59
Marquand, D. 72–73
Marsden, T. 197
Marston, S. 60
Martin, L.L. 80

Martin, S. 101–102, 105
Martinez, O.J. 237, 248–249
Marx, K. 65, 71–72
Masser, I. 7, 23, 39
Massey, D. 235
Mayes, D. 85, 87
Mead, W.R. 238
Miesenbock, K.J. 217
Molle, W. 82–86
Mønnesland, J. 10, 118
Morgan, K. 105
Munton, R. 202
Murdoch, J. 202
Murray, R. 59
Myhrman, J. 153
Nairn, T. 62
Nevalainen, P. 249
Nijkamp, P. 7, 19–20, 23, 34, 36, 50, 103
Nummela, J. 107
Oates, W.E. 75, 78–80, 92
Oksa, J. 10–11, 196
Okun, A. 82
Olofsson, C. 205
Olson, M. 75, 79–80, 92
Olsson, M-O 110
Paasi, A. 1, 12, 60, 107, 239, 241–246, 248–252
Paasivirta, J. 245
Palm, P. 212
Pepping, G. 39
Persson, L.O. 10, 112–113, 153, 157, 179
Peschel, K. 12, 125, 128
Philo, C. 105
Piore, M 64
Porter, M. 117, 207, 233
Pounds, N. G. J. 246
Pyy, I. 198
Qvortrup, L. 193
Ratti, R. 34
Reggiani, A. 39
Reichman, S. 34
Rietveld, P. 77
Rokkan, S. 5, 98, 103, 237–238, 243–244, 254

Rossera, F. 127–128
Rowthorn, R. 59
Ryynänen, A. 107
Räsänen, K. 230
Sabel, C. 64
Saegert, S. 62
Sahlins, M. 184
Sayer, S. 78, 80
Selstad, T. 104, 110–111
Senghaas, D. 123
Service, E.A. 184
Sharpe, L. 102, 108
Shields, R. 236, 254
Shils, E. 251
Sireni, M. 190
Sjøholt, P. 120, 130
˙Smith, A. D. 242
Smith, N. 7, 60–61, 65, 254
Snickars, F. 237, 254
Sogge, S. 107–108
Sørensen, O.J. 233
Spiekermann, K. 11
Stoker, G. 102
Storm Pedersen, J. 77
Storper, M. 11, 207
Strassoldo, R. 76, 236–237, 254
Stöhr, W. 106
Svidén, O. 14, 23, 39
Swyngedouw, E. 105
Söderström, L. 157
Sölvell, Ö. 208
Tapper, H. 178
Taylor, J. 82, 84, 86, 88
Taylor, P. J. 236, 245, 247, 254
Taylor, P. 60
Tegsjö, B. 112
Thrift, N. 105, 110–111
Timonen, S. 252
Tosi, A. 41
Tödtling, F. 106
Törnqvist, G. 128
Urwin, D.W. 5, 98, 103, 237–238, 243–244, 254
Varis, E. 249
Vartiainen, P. 10, 97, 108, 111
Vatne, E. 211, 215, 232–233

Veen, A. van der 7, 77–78
Veggeland, N. 77, 98, 107, 109–110
Ventura, F. 108
Vleugel, J. 39
Vloyantes, J. P. 247
Vuoristo, K.-V. 246
Väyrynen, R. 244, 247–248
Watermann, S. 243
Wegener, M. 11, 14, 39, 104
Westholm, E. 179
Whipp, R. 230
Wiberg, U. 113, 153
Wildasin, D.E. 78
Williams, C. 242
Wärneryd, O. 108
Ylä-Anttila, P. 233
Zander, I. 233
Östhol, A. 107

Subject Index

Accessibility 6, 11, 13, 116–118, 123, 159, 163, 211, 228, 231
Agglomeration 135–136, 230
Autonomy 19–20, 23, 45, 49, 51–52, 56, 102–103, 107, 195
Baltic Sea 108, 110
Baltic states 81–82, 247
Border 2–3, 5, 7–8, 12, 19, 22, 24, 26, 34–35, 48, 50–52, 56, 70, 75–82, 84, 89–93, 104, 107, 109–110, 131, 134–135, 165, 186, 206, 235, 237–239, 241–245, 248–253, 255
Border area 35, 52, 235, 237–239, 241–243, 245, 248–251, 253, 255
Border region 2, 7–8, 12, 48, 50–51, 75–79, 81, 84, 89–92, 104, 242
Bottleneck 22, 25–26, 32, 34, 38, 41
Bottom-up 49, 106, 111, 123
Capital flows (import, movements) 115, 144, 239
Capitals 62–63, 65–68, 76, 201, 231
Centre versus periphery 1
Centre–periphery 3, 5, 7, 54–55, 237, 245, 254–255
Centre–periphery system 254–255
Commission 26–27, 50, 84–85, 87–90, 101, 192, 208
Common agricultural policy (CAP) 200–201
Communications 13, 20, 23, 29, 61, 116, 123, 127, 132, 135, 153, 196–197, 206, 231, 250
Comparative advantage 83
Competition 7, 11–12, 22, 24–31, 34–35, 38, 41–43, 46, 50–51, 59–62, 66–68, 73, 78, 83–84, 103–104, 110, 128–129, 132–135, 147, 149, 153–154, 183, 189, 192, 206–207, 209, 215–216, 225
Competitive advantage 11, 21, 206–207, 209–210, 229
Competitiveness 25, 28, 46, 68, 71, 153–154, 177, 191–192, 206, 210, 212, 220, 228, 231

Congestion 32, 34, 36, 184
Continental Europe 2, 13, 134–136, 138, 141
Convergence 44, 84
Core–periphery 183, 185, 202
Co-operation 7, 12, 34, 41, 45–51, 53–56, 75–76, 78–79, 80–82, 89–92, 103–104, 107–110, 115, 129, 187, 194, 197–198, 200–201, 245, 247–248, 250
Cultural boundary 244, 252
Culture 3, 24, 44, 53, 62, 86, 98, 120, 129, 153, 160, 169, 173, 177, 186, 190, 192, 195, 197, 200, 202, 206, 238–239, 244, 250, 252
Dependency 1, 5–7, 20, 77, 97, 210
Depopulation 138–139
Deregulation 34, 152, 161, 166–167
Disintegration 18, 22, 24, 69
Disparities 1, 5, 41, 55, 76, 84, 86–88, 98, 103, 106, 189
Disposable income 157, 165–166, 176
Distances 6, 24, 91, 116, 124–128, 132, 135–136, 149, 211, 251
Distribution systems 156, 160
Division of labour 168, 186–187, 205, 209, 221
Dynamic competition 12, 41
EBRD 23
Economic growth 9, 12, 91, 123, 146, 153–154, 158, 176, 186, 191, 202, 228, 248
Economic integration 17–18, 20, 34, 44, 55, 68–69, 83–85, 131, 134
Economic space 44, 51, 55
Economic structure 190, 228, 230
Economies of scale 3, 34, 41, 51, 78–79, 84, 89, 205, 207, 215
EEA 131–133, 136, 149, 206
Efficiency 24, 29, 33, 38, 43, 46, 48, 56, 76, 79, 81–84, 88, 90–91, 93, 101, 157, 163, 177
EFTA 19, 131, 136, 187
Emigration 141

Subject Index

Employment 98–99, 103, 110, 129, 139–144, 147–148, 155–156, 160, 164, 170–171, 173–175, 177–178, 188–189, 212, 221, 241
Endogenous development 7, 41–42
Entrepreneurship 41, 53, 105, 153, 205
Environmental 21, 23–24, 28, 32, 50, 68, 72, 75, 77, 91, 101, 130, 148–149, 161, 197
Environmental policy (protection) 75, 77, 91, 101
Equity 24, 33, 47, 56, 76, 82–83, 88, 90–91, 93, 176–177
Europe of regions 7, 10, 35, 43, 45–46, 97, 101, 103–104, 106, 108–111, 131, 133–134
European Economic Space 44, 51, 55
European integration 6, 44–45, 55, 73, 75, 77–78, 83, 92, 97–98, 102–103, 115, 123, 130, 133–135, 149, 178–179, 185, 197, 199
European space 23, 28
European Union (EU) 2, 8–10, 12, 24, 35, 41–43, 48, 50, 54–56, 59, 72–73, 75, 81–83, 85, 89–90, 92–93, 101–103, 115–116, 118, 123–124, 127–128, 130–136, 138–139, 144–147, 149–150, 155, 178–179, 186, 190–192, 199–201, 206–207, 209, 225, 239
European unity 68–70, 72–73
Exploitation 71, 120, 122, 208, 229
Export orientation 153, 220, 222
External borders 8, 92
External effects 78–80, 82, 89–91
External resources 206, 210, 216–217, 222, 224, 226–228
Externalities 21, 23, 32, 38, 79–80
Federalism 41, 45, 47–48, 56, 75, 78–79, 81, 89, 92
Federation 46, 106, 186
Fiscal federalism 75, 78–79, 89, 92
Flexible production 45–46
Foreign direct investment (FDI) 115
Fortress Europe 19, 22
Fragmentation 69, 71

Free trade 9, 18–19, 46, 66, 84, 135, 187
Friction of distance 126
Frontier 26, 35, 51, 76, 80, 108, 186, 243, 249, 251
Functional integration 109, 111
Functional region 97, 103, 107–108, 210, 231
Gateway 34–35, 51, 77
GATT 64, 132
Geographical mobility 190
Geographical scale 7, 59–61, 63, 73
Globalization 17, 25–26, 30, 64, 70, 72, 105, 115, 206, 215
High-technology 162, 214, 216
Identity 17, 44, 61, 63, 70, 107–108, 134, 243–244, 246, 251
Identity regions 107
Ideological boundary 241, 246, 252
Immigration 23, 50, 52, 66, 165, 167
Income distribution 82, 157
Income transfers 9, 157, 165, 173, 209
Indigenous development 27, 110
Industrial clusters 208–209, 229, 231
Industrial structures 8, 88, 139, 141, 170, 229
Industrialisation 6, 8, 105, 122, 187–188, 196, 211
Information society 194
Information technology 29, 194, 196–197, 200
Infrastructure 6–7, 9, 11–13, 19–21, 23, 25–29, 32, 38, 42, 48, 51, 56, 61, 67, 75, 77, 79, 85, 91, 93, 105, 108, 110, 155, 162, 169, 173, 175–177, 179, 196, 199, 201, 209, 229, 231, 253
Innovation 11, 30, 36, 41–42, 48, 54, 116–117, 122–123, 184, 200, 207
Interaction 1, 12, 20, 22, 24–25, 46–47, 61–62, 115–116, 118, 123, 125–128, 197, 200, 237, 242, 246, 248–249, 253, 255
Intermodality 27, 31
Internal resources 210, 215, 217–218, 220–222, 225–227, 231

Subject Index 265

International trade 123–126, 129
Internationalization 62, 64–66, 152, 205–208, 210–211, 213, 215–218, 220, 223, 225–226, 228–232
Interoperability 26–27, 29, 31–32, 38
INTERREG 51, 90
Interregional co-operation 7, 45, 48–50, 53–56
Iron Curtain 238
Isolation 11, 90, 117–118, 193, 237
Joint ventures 23, 30
Just in time 25
Labour market policy 160–162, 164, 168
Labour market 82, 115, 138, 147, 151–153, 156, 160–162, 164–165, 168–173, 176, 190, 198, 210–211, 231
Legitimacy 49, 79, 103, 147, 153, 157
Less favoured region 86
Liberalisation 191
Local development blocs 197, 200
Local environment 23, 206–207, 211–212, 217–219, 223–224, 226–230
Local government 50, 56, 78–80, 89, 92, 106, 156, 163
Local milieu 11, 228
Local priorities 162
Localization 64, 105, 224, 226
Locational factors 207
Logic of collective action 75, 78, 80, 92
Logistics 25–26, 30
Low accessibility 116–118, 123
Low density 118, 132, 138
Maastricht Treaty 19, 26–27, 85, 89, 101, 131–132
Manufacturing industry 141, 143–144, 147, 153
Market economies 139, 146
Mass production 46
Mediterranean 19, 38, 51, 54, 118–119, 127
Migration 23, 64, 66, 116, 125, 148–149, 187, 189, 249

Missing links 19, 26, 29
Mosaic 101, 106
Multinational 18, 45, 65, 71, 105, 129, 186
National barriers 43–44, 50, 56, 133
Nationalism 44–47, 63, 69–71, 103, 239, 242
Nation-state 2, 4, 7, 17–19, 22, 24, 34, 43, 47–48, 50, 61–63, 65–66, 97, 102–103, 105–107, 131, 133–134, 237–238
Network 3–7, 9, 11–13, 17, 19–35, 38, 41–42, 44–45, 51, 53–56, 61, 73, 90–92, 103–105, 108–111, 115–116, 127, 143, 148, 185, 190, 192–193, 195, 197–202, 215, 223, 228–231
Network economy 17, 19, 34–35, 230
Network operators 19, 24, 28, 33
Network structure 3, 12
New Europe 23–24, 26, 55, 59–60, 62–63, 66–68, 70–71, 73
Niche-orientation 215
Non-metropolitan areas/regions 13, 207
Nordic countries 2, 8–12, 19, 81, 98, 106–108, 110, 115, 117–119, 122–124, 126–130, 132–136, 138–141, 144–146, 159, 186, 193, 205–210, 212, 214–215, 218, 225, 229–230, 238, 244–245
Nordic model 9, 159, 189, 209
Nordic peripheries 11–13, 97–99, 106, 110–111, 131, 138, 208–210, 230
Nordic Welfare State 186, 188, 190, 208
Outmigration 136–138, 148
Peripheral regions (periphery) 1–13, 20, 52, 54–56, 65, 69–73, 84, 92–93, 97–100, 103–104, 106, 108, 110–111, 117–120, 122, 128, 131, 133, 135, 138, 142, 147–149, 154, 160, 166, 169, 171, 178–179, 183–186, 189–192, 199–202, 206, 208–211, 215, 219–220, 226–227, 229–231, 235–239, 241, 243–248, 250, 252–255

Peripherality 1-2, 5-6, 9-13, 98, 103, 111, 115-118, 120, 122, 130, 138, 183, 185, 202, 235, 237-239, 241, 251-255
Peripheralization 237, 241-242, 246, 255
Physical barriers 20-21
Physical infrastructure 20
Physical planning 77-78, 93
Planning 24-26, 29, 48-49, 77-78, 93, 101, 108, 156, 158, 190, 192, 198, 216
Pollution 184
Population densities 13, 98, 135-137, 211, 241
Post fordist 97
Postnationalism 69-70, 72-73
Primary regional policy 167-169
Producer services 42, 120, 129, 225, 227
Productivity 32, 42, 85, 101, 120-121, 132, 141, 144, 147, 153, 157, 178, 187
Protectionism 19, 22, 118
Public good 92
Public policy 2, 7, 31, 82, 98, 144, 149
Public procurement 101
Public sector 5, 13, 23, 98-99, 105, 110, 119, 138, 144, 146-148, 151-154, 156-158, 160-161, 164-165, 169, 173-175, 177-178, 192, 194-195, 198, 200-201, 241
Public services 9, 13, 51, 141, 146-147, 151-153, 156, 163, 178, 188, 190, 192, 195, 200, 227
R & D 42, 104, 162, 199, 206, 225
Regional administration 43, 49, 53-54, 102, 104-108, 170
Regional autonomy 19-20, 23, 51-52, 102
Regional development 2, 10, 41-42, 55-56, 75, 79, 82, 86, 106-108, 110-111, 131, 144, 149, 154-155, 160, 167, 173, 177, 200-201
Regional disparities 5, 41, 55, 76, 84, 86, 88, 98, 103, 189

Regional effects 135
Regional equity 176-177
Regional planning 77, 108
Regional policy (policies) 2, 7, 9-11, 38, 41-43, 54, 56, 75-76, 82-83, 85-86, 88-91, 93, 100, 107-108, 133, 136, 138, 141-143, 149, 152-157, 159, 161, 167-169, 173, 176, 179, 188-189, 201, 249
Regionalism 45, 47, 56, 70-71, 102-103, 107
Regionalization 18, 47, 101-103, 108, 131, 135
Relative periphery 238
Remote regions 11, 13, 143, 147-149, 226, 229
Restructuring 8-9, 11, 13, 22, 35, 59, 60-61, 63, 73, 86, 88, 106, 136, 152, 183, 188, 191-192, 248
Rural periphery 186, 190-192, 199, 211, 220, 227
Rural policy 191-193, 198, 200
Rural population 187-188, 200
Scenario 36-38, 97, 103, 108, 111
Secondary regional policy 168-169
Semi-periphery 184, 236, 255
Single European Act 87
Single European Market (SEM) 44
Small and medium-sized enterprises (SME) 11, 32, 48, 205-230, 232
Social spatialisation 236
Socio-economic system 47
Soviet Union 12, 17-18, 69, 187, 241, 243-249, 251-253
Space 3, 13, 20, 23, 28-29, 44, 46, 51, 55, 60, 62, 105, 185-186, 193, 235-237, 239, 242, 244, 247-248, 253-255
Sparsely populated area (SPA) 155, 170-173, 198
Spatial environment 206-207, 210
Spatial implications 151, 159, 167-168, 178
Spatial mobility 23
Specialisation 6, 8, 44, 83, 105, 187, 205-206, 208, 215

Structural fund(s) 87, 89–90, 118, 136
Subnationalism 70
Subsidiarity 24, 45, 48, 50, 89, 101
Subsidies 19, 43, 50, 73, 87, 119,
   160–163, 166–167, 176
Sub–national administration 101–102
Sustainable development 33
Synergies 20, 24, 29, 32–33, 45–46,
   168, 176
Technological change 12, 207
Technology 20, 28–29, 31, 35–36, 121,
   155, 162, 193–197, 200, 205, 208,
   214–216, 224, 226
Telecommunication(s) 5, 13, 29, 31–32,
   116, 120, 123, 127, 153
Telecottage 193–196, 200–201
Telematic projects 193, 195
Telematic systems 193
Telematics 23, 31–32, 38, 195
Third Italy 35, 64, 105
Tourism 110, 127–129, 197, 250, 252
Trade 5, 8–9, 18–19, 24, 46, 59, 66,
   76, 83–84, 115, 117, 123–129,
   132–135, 139, 141–142, 183, 187,
   206, 209, 212–213, 246, 248–249
Trade barriers 133–134
Trade–blocks 25
Transboundary 91
Transfer programs 162, 169–170
Transition 11, 25, 81, 199
Transnational firms 45
Transnationalism 65
Transport costs 11, 85, 116, 124–125,
   129, 211, 215
Transport demand 26
Transport policy 26, 31
Trans–European networks 11, 27, 29
Unemployment 9, 82, 87–88, 98, 105,
   136, 138–139, 141, 146–147, 154,
   161–164, 166, 170, 178, 191, 239,
   249
Urbanisation 149, 187–188
Vertical integration 30, 51
Welfare state 9–11, 115, 123, 151–153,
   157–159, 162, 164, 174–175, 177,
   186, 188–191, 195, 198–201,
   208–209, 239

Contributors

Dirk-Jan Boot is Assistant Professor at the Department of Public Policy and Public Administration, University of Twente, the Netherlands. His current research interests are evaluation of regional policy instruments and cross-border cooperation.

Riccardo Cappellin, Professor of Economics at the University of Rome Tor Vergata, is Vice President of the European Regional Science Association. He is author of various publications on the issues of regional industrial policy, technological change, economics of producer services and European regional policy.

Heikki Eskelinen is a senior researcher at the Karelian Institute, University of Joensuu, Finland. His research work has focused on the structural economic problems of peripheral regions. He was chairman of the Local Organising Committee of the 1993 Advanced Summer Institute in Regional Science, which led to the initiative of publishing this book.

Stephen Fournier is a senior researcher at the Department of Regional Planning, Kungl Tekniska Högskolan, Stockholm. His major areas of interest are regional planning and regional economic change, particularly with regard to labour issues. His expertise is in the development and use of computer models for applied regional analyses, especially for policymakers.

Sven Illeris is Professor of Geography at Roskilde University, Denmark. He previously worked in Danish Government Administration, at the Local Government's Research Institute, Copenhagen, and with the European Commission on problems of regional development and regional policy, as well as on the geography of service activities.

Merja Kokkonen is presently the coordinator of the international Human Geography Studies at the University of Joensuu, Department of Geography and Regional Planning. She specializes in European spatial policy and regional development.

Leif Lindmark is Professor of Business Administration at Umeå Business School, Sweden. He has published a number of articles and books about regional variations in new firm formation, business dynamics and the internationalization of SMEs. A current area of research is the use of new information technology in SMEs and in regional development.

Jan Mønnesland works as a researcher at the Norwegian institute for urban and regional research (NIBR) in Oslo. His research covers macro regional development both on the domestic and European level. The regional effects of economic integration have been a major topic in a newly finished project within the ESF programme on regional redistribution in Europe.

Peter Nijkamp has a chair in regional economics at the Free University in Amsterdam. His research has a strong international orientation and is often characterized by a quantitative-analytical approach. He is involved in various European research initiatives and programmes.

Jukka Oksa, rural sociologist, is a researcher at the Karelian Institute of the University of Joensuu, Finland. His research interests cover rural transformation and rural policies in peripheral Europe and North-Western Russia.

Anssi Paasi is Professor of Geography at the University of Oulu, Finland. His research fields are the history of geographical thought, social theory, political and regional geography, and the problems of regionalism, regional consciousness and regional identity.

Lars Olof Persson is senior researcher at the Research Group on Regional Analysis, Department of Infrastructure and Planning, Kungl Tekniska Högskolan, Stockholm. He has been working in the fields of regional analysis and rural studies, previously at the Swedish Agricultural University and within the Swedish Ministry of Labour. He is currently involved in research on regional labour markets, mobility and regional policy.

Neil Smith is Professor of Geography at Rutgers University and Permanent Fellow at the Centre for Critical Analysis of Contemporary Culture. He has written and edited several books and many articles on uneven development, gentrification, questions of scale, and the politics and history of geography. He will spend 1995-96 on leave as a John Simon Guggenheim Fellow, completing a study of "Isaiah Bowman and the geographical pivot of history".

Folke Snickars has been Professor of Regional Planning since 1985. He was the President of the European Regional Science Association during the period 1989-1994 and is now the European Editor of the Papers in Regional Science. He is currently Chairman of the Department of Infrastructure and Planning, Kungl Tekniska Högskolan, Stockholm. His research interests include transport, communications and regional development, planning and forecasting models of regional analysis, decision support systems in planning, and planning negotiations.

Perttu Vartiainen is Professor of Human Geography at the University of Joensuu. He has previously been Professor of Applied Geography for Planning at the University of Helsinki. His current research fields are urban systems change and spatial development policy.

Anne van der Veen is Associate Professor at the Department of Public Policy and Public Administration, University of Twente, the Netherlands. His current work interests are fiscal federalism, European regional policy, evaluation of regional policy instruments and cross-border cooperation.

Advances in Spatial and Network Economics

B. Johansson, C. Karlsson, L. Westin (Eds.)

Patterns of a Network Economy

1994. VIII, 314 pp. 33 figs., 44 tabs. Hardcover DM 148,-
ISBN 3-540-57824-2

A.E. Andersson, D.F. Batten, K. Kobayashi, K. Yoshikawa (Eds.)

The Cosmo-Creative Society

Logistical Networks in a Dynamic Economy

1993. VIII, 296 pp. 67 figs., 31 tabs. Hardcover DM 148,-
ISBN 3-540-57158-2

A. J. Reynolds-Feighan

The Effects of Deregulation on U.S. Air Neworks

1992. XIV, 131 pp. 15 figs., 30 tabs. Hardcover DM 88,-
ISBN 3-540-54758-4

M. J. Beckmann, T. Puu

Spatial Structures

1990. IX, 139 pp. 40 figs. Hardcover DM 68,-
ISBN 3-540-51957-2

Terms subject to alteration

B.M. Roehner

Theory of Markets

Trade and Space-time Patterns of Price Fluctuations. A Study in Analytical Economics

1995. XVIII, 405 pp. 149 figs. Hardcover DM 175,- ISBN 3-540-58815-9

The purpose of the book is to investigate the foundations of international and interregional trade at the microeconomic level of spatially separated commodity markets. At this level, price arbitrage and local disparities in production and demand functions are the main determinants. The model, referred to as the Enke-Samuelson model, is developed step by step.

A. Sen, T.E. Smith

Gravity Models of Spatial Interaction Behavior

1995. XVI, 572 pp. 13 figs., 28 tabs. Hardcover DM 178,- ISBN 3-540-60026-4

This book presents an up-to-date, consistent and unified approach to the theory, methods and applications of the gravity model – which spans from the axiomatic foundations of such models all the way to practical hints for their use. "I have found no better general method for use in applied research dealing with spatial interaction... It is against this background that the present book by Sen and Smith is most welcomed."

Walter Isard

Springer

Tm.BA95.03.23